TS183 .L45 2005
0134109656808
Lembersky, Michael

Moder
tech

MW01518310

GEORGIAN COLLEGE LIBRARY 2501

25 01 01(00)
$36.25

MODERN MANUFACTURING TECHNOLOGY
&
COST ESTIMATION

A systematic approach with engineering vision

by

Michael Lembersky
and
Lana Lembersky

Learning Resource Centre
Georgian College
One Georgian Drive
Barrie, ON
L4M 3X9

authorHOUSE™
1663 LIBERTY DRIVE, SUITE 200
BLOOMINGTON, INDIANA 47403
(800) 839-8640
WWW.AUTHORHOUSE.COM

© 2005 Michael Lembersky and Lana Lembersky. All Rights Reserved.

No part of this book may be reproduced, stored in a retrieval system, or transmitted by any means without the written permission of the author.

First published by AuthorHouse 07/15/05

ISBN: 1-4208-6870-5 (sc)
ISBN: 1-4208-6869-1 (dj)

Library of Congress Control Number: 2005905772

Printed in the United States of America
Bloomington, Indiana

This book is printed on acid-free paper.

Cover designed by Natalya Lembersky.

We dedicate the book to OUR WONDERFUL DAUGHTERS,

MARGARITA AND NATALYA

Table of Contents

INTRODUCTION & OVERVIEW

We have entered a new era of fundamental change in manufacturing technologies. Knowledge, experience, and skills in combination lead to action; they also are a prerequisite for the successful practitioner. He must not only be proficient at using technology, but he must also retain that proficiency as the technology changes. On the other hand, knowledge, people, and machines are more costly. The most effective company strategy can be developed based on accurate cost information. It is very important to recognize the role of a PRODUCT COST SYSTEM in the successful business process. Cost modeling is an integral component of a companywide cost system. The book *MODERN MANUFACTURING TECHNOLOGY AND COST ESTIMATION,* in a reader-friendly form, contains information, concepts, and ideas for experts in the areas of technological processes applications, cost estimation, and analysis. The objective of our book is to provide new information and refresh previous knowledge, as well as to serve as a concise and useful source of up-to-date information for practicing engineers and managers. Furthermore, this book can be of service in cases when a professional deals with limited information about product specifications. We also hope that this book will help to allocate knowledge for productive use. The key is a systematic approach with engineering vision to the best practices that focus on reducing cost in a changing environment. Thus, you will be able to uncover new business opportunities and negotiate better deals.

The book's broad coverage should be especially useful to a busy professional who will not have time initially to research all of these topics in-depth, but who requires an immediate working knowledge of a process. People say: "A picture is worth a thousand words." The extensive set of illustrations, tables, and diagrams visually supports the thought process. The book (see Fig. 2) encompasses advanced manufacturing technology, knowledge-based applications, and cost-reduction methodology (CRM). Well-organized cost systems allow management to effectively analyze how market conditions will influence the company's economic situation. Ultimately, misleading cost data can direct management to the wrong or ineffective product solutions. The following scenarios may result, when pricing policies are based on misleading cost information: wrong products from a

3

margin point of view are heavily promoted in marketing campaigns. Decisions based on available product cost data have a profound effect on the company. The well-organized and integrated cost system ultimately will allow the professional to meet business performance objectives and to avoid or minimize the risk.

A customer decided to have a product with better functions and features, which will require new material and manufacturing process and equipment; a cost-responsible specialist develops the cost model and provides the product cost and tooling costs. Based on the cost information, finance professionals develop financial performance parameters and the manager makes a decision. The chain of events begins with the cost estimation process. Cost projections must be accurately calculated and presented to management, then product price analysis will be effectively conducted on different levels. In recent years, complexity of products has increased substantially, as well as variability of manufacturing processes. This requires management to pay more attention to PRODUCT COST SYSTEM improvements within the company. Managers add more people to the costing process and encourage them to use new software tools dedicated to cost estimation. Cost saving justification for product and process improvement is required in order to obtain funding authorization.

Every day, new people receive training on cost analytical tools and assignments based on the variety of products manufactured. They are required to keep consistency across all types of cost evaluation studies like never before. It is important to utilize new information. The use of Web-based technologies is becoming increasingly common. The Internet nowadays is a place where a practitioner can access viable sources for virtually every subject; a number of Web sites have advanced features and data. Each of these Web sites can provide rich information content to the practitioner. The information is typically targeted to a narrow technological niche, such as machining technology, stamping dies and materials, etc. Some Web sites allow you to use them as a single source to generate a cost estimate; others are just an additional source. The key point is that the Web site replaces several reference sources, as well as providing the most updated information, which cannot be found anywhere else. This book offers selected Web sites that will be helpful for busy people in the process of cost development.

This book is the attempt to reduce the gap by placing in one published source, information about modern technological knowledge and about cost justification methodology. In each chapter, the text is accompanied by tables and diagrams. We would like this book to be the extensive source for business professionals, purchasing specialists, engineers, and managers. The intent of the book is also to serve as refresher for the practitioner who is already aware of some information that is in the book, but would like to refresh his knowledge. We would be happy if after reading this book, the professional would be able to make faster, more confident technological, costing, and sourcing decisions. The book may be the source for providing cost reduction training (CRT) for cost estimators, cross-functional team (Fig. 1) members, and purchasing professionals, as well as engineers.

The book (Fig. 2) is going to be a valuable tool continuously because it includes a wealth of additional published sources. Getting publications in a team communication loop allows the reader to utilize some specialized sources (books, periodicals, Web sites) in order to improve technological knowledge and expertise in addition to training. Whether you are doing research or just looking for some information, the bibliography can help you to find relevant publications pertaining to the subject area of manufacturing technology. The recommended literature is also going to help you in facilitating the training process.

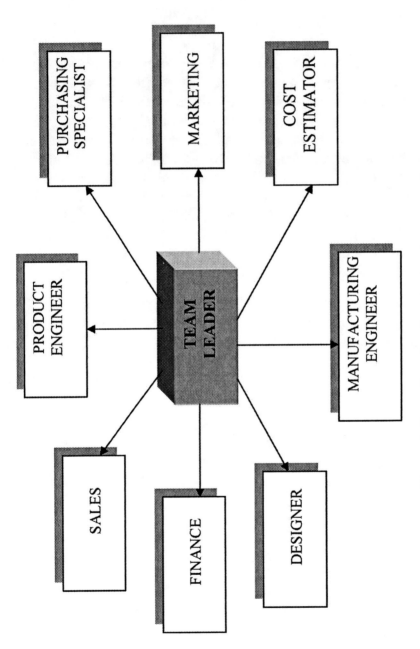

Figure 1. Cross-functional team structure example.

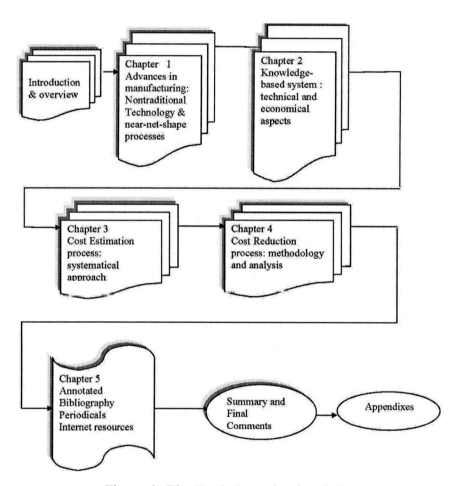

Figure 2. The Book Organizational Structure

Accurate cost data help to proactively manage the production process and support competitive cost solutions. A monitoring element should be considered as a part of PCS which represents business-oriented cost evaluation methodology with the following functions:

- New product cost development

- Tooling cost development

- Engineering changes cost estimation

- Supplier selection and evaluation assistance

- Alternative cost analysis studies and projections

- Evaluation support of investment analysis

- Technological innovations and feasibility studies

- Product competitiveness cost analysis

- Cost tracking and reporting activities

The main structural components of the cost reduction process (CRP) are COST ESTIMATION, VALUE ANALYSIS, and TECHNOLOGICAL KNOWLEDGE. These are drivers of the cost reduction process. This book is an attempt to provide a systematic approach to the PCS analysis, including materials and manufacturing processes coverage. There are many types of industries and products. However, common MANUFACTURING PROCESSES AND MATERIALS are applied to the entire product variety of many different industrial products.

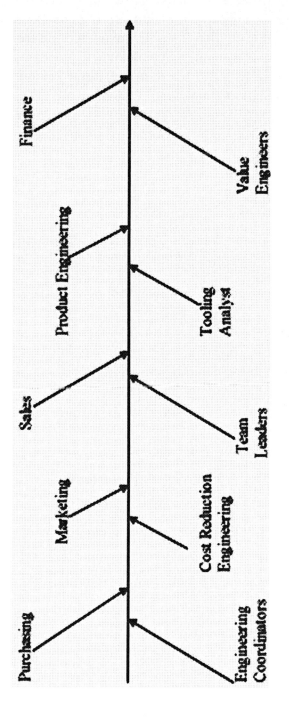

Figure 3. The fishbone diagram of cost data users

As you can see, different products are manufactured from the same type of materials, using similar manufacturing processes. However, As you can see, different products are manufactured from the same type of materials, using similar manufacturing processes. However, capable of effectively supporting business decisions, great attention should be devoted to the following PCS segments:

- Cost estimating methodology

- Cost models development (work breakdown structure, bills of materials)

- Cost drivers analysis and calculations

- Best practices database: materials, manufacturing processes, equipment, tooling

- Spreadsheet analysis and cost evaluation software

- Consolidated reference resources

Only when these segments are properly maintained and updated is the company capable of building a profitable and successful pricing strategy. The fishbone diagram (Fig. 3) illustrates the users of cost data across the company departments. When PCS is properly monitoring the consistency of cost data, then cost reduction strategies are developed in the most efficient way.

It is also important to consider technical feasibility, which refers to the ability of processes to take advantage of the current state of the technology and apply to continuous improvement. This book offers a variety of traditional and nontraditional manufacturing processes and materials, their description and comparative analysis. Nontraditional technologies are represented by electrochemical machining, electrical discharge machining, and near-net-shape processes. New materials for products, as well as for dies and molds, require applications of processes that would be able to achieve tight tolerances and fine finished surfaces. The other side of advanced process knowledge is the ability to conduct the alternative cost study with the optimal economical solution based on detailed product analysis. The intent was also to assist management in the training process. The information presented in this book could be used for training purposes. The practitioner will be able to effectively contribute to product development, business planning, or financial feasibility analysis. Purchasing specialists should use this book in supplier selection, quotation review, and sourcing process. Cost reduction engineers, team leaders, business coordinators, sales representatives, marketing and account managers will be able to make better-informed decisions in relation to business projects or proposals and their economic feasibility.

CHAPTER 1

Advances in manufacturing: nontraditional technology & near-net-shape processes

Progress in manufacturing has been spectacular in terms of productivity, new process developments, and implementations, as well as multiple operations concentrated on a single machine. Today's manufacturing industry is facing challenges from advanced difficult-to-machine materials (alloys, composites), and high design requirements (high-precision, complex shapes, high surface quality, manufacturing costs). Advanced materials play an increasingly important role in modern manufacturing industries. Only consideration of complex interactions of parameters across various layers in the manufacturing process enables a new technology to become an acceptable replacement. That is why we also walk the reader through some nuances of technological developments in this chapter.

The thermal, chemical, and mechanical properties of materials (corrosion resistance, wear resistance, strength, heat resistance) have significantly improved, which makes traditional processes unable to machine them or unable to machine them economically. This is because usually machining is based on removing material using tools harder than workpieces. For example, the classical method of grinding is suitable only to limited cutting materials. Sometimes the cost of machining hard materials exceeds 50 percent of the total part cost. There are also design requirements such as more complex shapes, complex cavities in molds and dies, non-circular, small, and curved holes, low-rigidity structure of parts and components. Traditional machining is ineffective for these applications. To meet these challenges, nontraditional processes need to be applied.

To get a clear understanding of various options of new processes, we need to analyze their capabilities. This chapter is a direct response to the ever-increasing need for knowledge of modern process developments and their impact on cost. We hope that this chapter will expand your thought process, and more ideas will deliver a measurable improvement to your company's business process. The knowledge of technological advances is essential. The comprehensive coverage of modern manufacturing processes will be able to provide multidimensional views on economic problems. Manufacturing engineers, cost reduction engineers, cost estimators, project engineers, purchasing specialists, sales and marketing specialists—including the team leader of a cross-functional team—bring different outlooks and expectations to the cost reduction process. They do not think alike and have diverse backgrounds and experience. This chapter takes the reader through the technological developments that play an important role in cost reduction efforts. The knowledge subsystem would help to use a common "technical language" for representing their opinions and views. Besides that, in many cases, parameters for analyzing different processes are not common. Feed rate and revolutions per minute (RPM) apply to machining, while strokes per minute applies to stamping. Preliminary calculations for baseline estimates are based specifically on process-related data. In addition to that, manufacturing processes are based on knowledge of science or previous technology to certain useful objectives. That is why they are characterized by different parameters. Laser cutting is based on physical principles other than electrochemical machining. Both are fundamentally different methods, while cold forging and hot forging both are types of forging applied to different objectives.

Nontraditional technology: features & capabilities

In this chapter, we want to concentrate our attention on nontraditional processes. Advanced technology is a catalyst to the company's growth. One way to classify the entire group of manufacturing technologies is to divide them into two major subgroups: traditional and nontraditional technology. Nontraditional technology can be classified into the following segments: mechanical and non-mechanical segments. The mechanical segment of processes (MP) represents material removal and forming methods based on mechanical interaction of the cutting edge with the workpiece materials. The non-mechanical segment of processes (NMP) is a wide range of processes based on physical and chemical principles: electricity, magnetism, electrolysis, heat effect, etc. Integration of these principles enables the user to remove material or shape a part with fine surface finish and tight tolerances. Enhancing conventional technologies, however, based on electricity, magnetism, electrolysis, or heat effect allows the user to increase productivity, reduce cycle time and cost accordingly. Advances in computer control have increased operation efficiency and productivity. Since its introduction, the precision, speed, and capabilities of NMP have grown tremendously. In addition to the tooling niche, it expanded to encompass virtually all areas of electrical conductive material machining, forming, and finishing. NMP were enhanced by CNC systems, robots, palletization, automatic toolchangers, and slug removers. The first models have been adapted to regular machine tools, such as turning machines, drill presses, grinding and milling machines.

The reason for the rapid advancement in material removal rates, high forming speed or plating productivity is the continuing evolution of the power supplies (PS). A wide range of process parameters (current density, quality of finish surface, etc.) depend on PS characteristics. Another important component of NMP is electrodes. In order to reduce electrode wear, engineers try to optimize design, material selection, and manufacturability. Software modeling programs play a substantial role in electrode design, shape prediction, and process modeling, which reduces cost. Modern CNC systems are able to support fine incremental movements, feed rates, and voltage, and provide immediate response to changes in cutting conditions. Vertical type sizes range from small benchtop models to huge machines. The workpiece dimensions vary from fender to hood die. The appealing feature of NMP machines is that they can run overnight or over a weekend unattended. They can be incorporated in unattended systems using palletization and robotics. Some production lines are equipped with vision systems so the robot can see the component orientation and execute a pickup and transfer. An automatic quick toolchanger is widely used in regular or unattended situations. This method enables the user to apply multiple electrodes and change them as needed. Now we know what makes NMP an attractive alternative to traditional mechanical processes.

One of NMP's processes is electrochemical machining (ECM) (Fig. 4). This method can be described as a combination of electrochemical process with machining operational function. In other words, to remove material, we apply electrochemical dissolution. So ECM cycle time depends on electrochemical and machining parameters.

At the same time, the fundamental view of the manufacturing technology helps in finding ways to apply it efficiently with respect to cost reduction.

Usually, ECM uses 500-20,000 A, typically and low-voltage (10-30V) DC power and high electrolyte flow rate through the machining gap (3,000-6,000 cm/sec). Electrolysis is the chemical process that occurs when current is passed between two electrodes dipped into electrolytical liquid solution. Typical applications of electrolysis are electroplating and electropolishing. In the case of ECM, the electrolyte is forced to flow through the interelectrode gap with high velocity, approximately 5m/s and more, to intensify the mass/charge transfer through the anode and to remove dissolution products, e.g. hydroxide of metal, heat and gas bubbles generated in the gap. One of the primary functions of the electrolyte is to carry away the heat generated by ECM and to maintain a constant temperature in the machining zone. The tool is moving toward the workpiece while maintaining a small gap. Electrical potential difference is applied across the electrodes. The feed rate range is from 0.1 to 20 mm/min. Since there is no contact between the tool and the workpiece, ECM is able to machine thin-walled, easily deformable parts, and also brittle materials. No residual stress remains after ECM. In past years, mathematical equations have been formulated for describing tool design process.

There are some important advantages of ECM: the high speed of the process, for example a 15,000 A machine is able to remove material with material rate removal (MRR) of 25 cm^3/min. ECM is well known as a no-tool-wear process, with excellent finished surface conditions, and high accuracy. ECM is especially useful in the case of complex shape and hard-to-machine components, including heat-treated parts. ECM is applicable for casting tooling, molds, and dies. Tool steels, used for tooling components, are hard-to-machine materials, and ECM better than traditional machining methods, enables the user to achieve tight tolerances, great finish with high productivity, which also differentiates ECM from other nontraditional operations. Among ECM operations are: electrochemical grinding, electrochemical polishing, electrochemical milling, electrochemical deburring, electrochemical broaching, electrochemical shaping of rotating workpieces, and electrical discharge machining in combination with electrochemical machining in one operation which is called ECDM. The capability of ECM of smoothing irregularities after electrical erosion increases the efficiency of the ECDM operation. In the case of long-channel ECM machining instead of drilling, the company has the opportunity to eliminate about eight to ten operations (preliminary, semi-finish, and finishing operations) depending on length, surface finish, and tolerances. Sometimes the channel is so long that traditional drilling cannot be applied. Electrochemical broaching is applied to making splines and gears. Electrochemical grinding using a conductive bonded abrasive tool is particularly effective for the machining of difficult-to-cut materials such as sintered carbides, metallic composites, and titanium alloys. Material is removed through a combination of electrochemical action and traditional mechanical grinding. Approximately 90 percent of the material is removed through electrochemical action and only 10 percent by mechanical grinding. Because only a small amount of material is removed by grinding action, the wheel life is typically ten times longer than the life of a conventional grinding wheel. The gap produced by ECM reduces contact arc. Considerable reduction in cutting force significantly decreases abrasive material wear. That means ECM grinding provides machining cost savings when using expensive grinding wheels. ECM grinding can be applied to high-complexity parts in terms of shape. A typical ECM machining system has four major subsystems:

- The machine

- The power supply

- The electrolyte circulation system

- The control system

Basic operating parameters of ECM are:

- Workholding voltage between the tool electrode (cathode) and workpiece (anode)

- Machining feed rate

- Inlet and outlet pressure of electrolyte (flow rate)

- Inlet temperature of electrolyte

The value of current used in ECM is dependent on the above parameters and dimensions of the machining surface. For manufacturing results of ECM, the distribution of current density on the anode is very important. Electrochemical properties of workpiece material and electrolyte play an important role. Typically the following electrolytes are used for ECM:

- Most used: NaCl (sodium chloride) at 60 to 240 g/l

- Frequently used: NaNO3 (sodium nitrate) at 120 to 480 g/l

- Temperature: 68 to 122° F

- Velocity: 5,000 to 10,000 fpm

- Inlet pressure: 0.15 to 3 MPa (22 to 436 psi)

- Outlet pressure: 0.1 to 0.3 MPa (15 to 43.6 psi)

- Electrode material: brass, copper, bronze

The ECM capability of any material can be described by electrochemical machinability coefficient (EMK). As compared with electrical discharge machining (EDM), ECM has no electrode wear, while EDM electrode wear increases with high precision requirements. We think would be very helpful to provide an example of current calculation for electrochemical machining. Let's say we need to determine what current is required to achieve a material removal rate (MRR) of 12 cm³/min.

Calculations:

1 min. = 60 sec.

MRR = 12 cm³/min. = 12 cm³ / 60 = 0.2 cm³/sec

MRR = 0.2 cm³/sec

A = 56 g (periodic chemical table)

D = 7.8 g/cm³ (density)

Z = 2 (from chemical reaction)

F = 96,500 Coulombs (Faraday number)

I = ? (electrical current).

MRR = 56 I / 7.8 * 2* 96,500

I=MRR*7.8*2*96,500/56=0.2*7.8*2*96,500/56=5376 A

Actual rates can vary from ideal process conditions, because real process conditions come to effect. When conducting cost-benefit analysis, it is important to take into consideration the labor factor. In the case of ECM, labor requirements are lower than in traditional mechanical processes (turning, milling, grinding). The following is an example of ECM forging die sinking:

Die sinking operation	Labor requirements
Milling machine	450 hrs
Rough milling	30 hrs
EDM	45 hrs
ECM	20 hrs

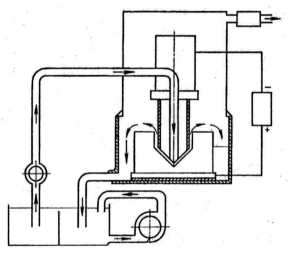

Figure 4. Illustration of Electrochemical Machining

As we can see, in this particular case, ECM is a more competitive process. Oftentimes, full-blown economic justification is required in order to make a decision. ECM is applied also, where traditional manufacturing technologies are not able to perform an operation. An example is electrochemical drilling of a small diameter deep hole. There is no other way of machining this type of hole, and no direct comparison of labor requirements can be made. Another example is when parts are traditionally machined separately, and then installed and assembled. ECM enables the user to perform this type of assembly from a solid blank. This is a very economically attractive approach to reduce cost up to 50 percent.

One of the advanced non-mechanical processes is electrical discharge machining (EDM). EDM has been developed based on a problem with no connection to machining. The engineers were studying erosion of electrical contacts. The rate of erosion was greater when the contacts were submerged in oil. It was decided to apply the erosion effect to the metal removal operation. Thus, the problem became a method of nontraditional machining. Since then, EDM has been implemented in many companies as the process for a wide variety of hard-to-machine materials and complex geometrical shapes, tight tolerances, and fine surface finishes. Furthermore, EDM enhanced with CNC systems is a highly competitive method for making forging dies, casting tooling, plastic injection molds, and tooling for powder metals. It enables the user to machine simultaneously multiple highly precise parts from a single clamping. CNC systems continuously monitor EDM operations, preventing short circuits, which improves process efficiency. The volume of metal removed varies with the melting point of the workpiece. The EDM operator is able to work with a number of machines, because of the increased autonomy of EDM. Since everything is operated independently, he can perform any other tasks while the machines are running. The EDM process depends to a large extent upon the electrode materials selected. However, various electrode materials perform differently when used with different types of power supplies. The EDM multi-axis center allows the user to machine multiple cavities in a single piece of tooling. EDM planetary function is a sequence of orbital tool movements that can be used for generating circular and angular profiles, spheres and helical curves in workpieces, applying economically effective electrodes. EDM can be successfully implemented in a moldmaking and diemaking environment. One of the EDM operations is called electrical discharge wire cutting (EDWC). The small-diameter wire is a tool and travels from the supply spool through the workpiece to a take-up spool. EDWC uses thermal energy of a spark to remove material. Electrode wire may be made of brass, copper, tungsten, or molybdenum. The diameter of the wire may vary from 0.003 to 0.012 in. depending on the desired kerf width. Basically, EDWC works like a band saw, except EDWC is more precise and cuts a narrower kerf. Contours can be tightly controlled and extremely sharp angles can be cut. Hardness and toughness of the metal do not affect the machining rate, so EDWC is often used to machine heat-treated metals or sintered carbides. The workpiece must be electrically conducted. EDWC permits producing mating male and female parts from separate blanks of material using the same CNC program. EDWC machines are also suitable for making some conventional EDM electrodes. Punches, dies, and stripper plates can be cut in very hard metals. Tapers can also be cut by shifting the working plane. This makes it possible to produce both the punch and die at the same time.

The next type of EDM application is electrical discharge grinding (EDG), which removes material the same way as EDM and uses a rotating wheel. One of the applications is grinding thin sections without distortion. The following essential EDM process components are:

- Electrode materials: graphite, copper, etc.

- Process variations: EDM die-sinking, ED grinding, wire EDM, deep hole drilling (L/D 20:1).

- Machines are equipped with CNC systems.

- Workpiece material: electrically conductive material. Hardness does not affect EDM process.

- High degree of automation, burr-free parts, large or small holes.

- Workpiece complexity: capable of EDM machining high complexity parts with undercuts.

- Typical applications: dies for forging, casting, extrusions, blanking, etc.

Currently, there are many product applications that EDM can do better than any other machining method, so they must now be done by EDM in order to be competitive. EDM applications also increased profitability of tooling operations. Highly competitive companies in the tooling business are the ones that implemented EDM for making molds and dies.

When a process engineer generates an EDM operation instruction for a particular part, it should include the following:

- Type of parts

- Overall part dimensions (length x width x height)

- Number of electrodes

- Number of surface passes

- Machining allowance (material to be removed)

- Condition of material (hardened, annealed, and so on)

- Tolerances, surface finish

- Machine tool model

- Special purpose tooling and fixturing

The choice of a machining scheme and machine type depends on the part's geometry and the depth of the cut (as in drilling or milling). Now, it is also important to mention about design for manufacturability in relation to EDM. When EDM manufacturability is considered, it enables the user to reduce operation cost compared with traditional design for manufacturability. The concurrent engineering approach to the process is a substantial saving factor in manufacturing cost.

The combinations of several individual processes are called cross innovative processes (CIP). The reasons for implementation of CIP are to make use of mutually enhanced process advantages. The performance parameters of CIP differ from individual processes. When we combine in one operation electrochemical (ECM) and electrical discharge machining (EDM), physical and chemical conditions of the CIP are significantly different from individually applied processes. Productivity of combined CIP process is three to six times greater than productivity using individual processes – ECM and EDM separately. Generally speaking, there are two types of CIP:

1. Processes involved in CIP directly participate in material removal or forming.

2. Only one participating process directly removes material while others only assist in removal by changing the operation conditions.

Examples: electrochemical grinding, and EDM with ultrasonic assistance (EDMU). The beneficial effects from EDMU are as follows: high-frequency motion of the electrode and workpiece creates better ejection of the molten material. This increases the removal rate, improves productivity, and can be successfully applied to machining slots and grooves. One more application is laser-enhanced ECM. The laser helps to better localize the electrochemical dissolution during the process, which increases the accuracy and the removal rate, and improves surface quality.

Cost reduction directions in ECM are correlated with the usage of economical electrolyte, cathode material, and reasonable tolerances. The pre-shape workpiece roughly conforms the tool, that improves current density balance and reduces removed material amount, resulting in shorter cycle time. The optimization of surface finish parameters enables the user to save money on every part that does not require a secondary processing step to correct defects or improve surface finish. It is recommended to select workpiece material with high material removal rate which contributes to productivity increase and operation efficiency. As shown above, part manufacturability analysis for ECM allows the user to identify the most economical factors and justify the application of ECM. Why is that important? Because there are a number of ways available to efficiently accomplish manufacturing operations. For example, drilling a hole can be performed either by a traditional method using a twist drill or by any of the following methods: ECM, EDM, laser machining, or electron beam machining. Laser machining can be improved with water jet laser guidance of the beam, which allows the user to achieve a cutting speed of more than 4 m/sec.; application of a water-jet guided laser improves regular laser machining operation. Continuous improvement of applications, knowledge, and comparative cost justification, however, is always required in order to make the best business and technical decision.

Near-Net-Shape Processes

There is group of processes which is called near-net-shape (NNS) technologies: forming, casting, forging, powder metallurgy, plastic injection molding, etc. One of the advanced types of forming is electromagnetic forming (EMF) (Fig. 5). The EMF production rate is between 900 parts/hr. and 2,400 parts/hr. In other words, EMF is the high-speed process which allows forming metal parts over plastics and composites. The machines for EMF are efficient and highly productive. In many cases, parts can be formed in a single operation with significant cost savings. The benefits of EMF over conventional forming are as follows:

- High-precision metal parts forming over plastic and composites

- No contact with the workpiece, no lubricant required

- The machine are energy efficient, easily installed

- Uniformed parts, no tool wear

- No heat distortion

EMF applications include forming and assembly operations as well as replacing swaging, spin rolling, soldering, and welding. EMF also allows fasteners elimination. The nature of EMF makes it highly suitable for automation. The EMF workcell can provide automatic part orientation. That means significant manufacturing cost reduction is achievable by decreasing cycle time and improving quality with EMF.

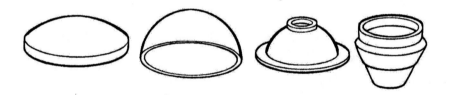

Figure 5. Electromagnetic forming applications

Orbital forging represents another NNS process. This is a significant technological breakthrough in cold-forming operations. It produces NNS parts; no secondary operations are required, which allows the user to save money. Presses for orbital forging are completely automated, equipped with robots. The change time from one part to another is very short and can be achieved with hydraulic clamping tooling. Production rates are from ten to twenty parts per minute. Among the advantages of the orbiting type of cold forging are: short cycle time, reduced material requirements, increased material strength, tight tolerances, and high surface quality. In the case of large size components, die cost and press tonnage requirement for orbital forging is less than with conventional forging. Orbital forging is able to produce parts and components with more complicated shape than other types of forging. A forging press can be added in the manufacturing cell in line with machine tools as well, which saves floor space and allows for better process control. Cold forging technologies are suitable for mass production; because of long tool life and small tool wear, cold forged products can be produced with tight tolerances, fine surface quality, improved mechanical properties, and less machining operations required.

As indicated previously, combinations of the individual methods will lead to optimized processes and thus to an increased components cost effectiveness and competitiveness. Application of advanced processes can substantially increase cost savings. We need to keep in mind that in many situations, innovative technology selection is more efficient than just incremental solution for cost reduction. The benefit of innovative technology is high, especially when long-term decision is required.

One of the near-net-shape technologies is casting. New developments in casting enabled the user to improve the process efficiency in terms of productivity and quality. High-volume production casting especially is very competitive among other manufacturing technologies. Casting also can be implemented in manufacturing cells with machine tools, when machining is required as a secondary operation. It is recommended to analyze stampings and welding subassemblies in order to investigate the opportunities to implement casting, which could save material and decrease parts count. Three-piece steel weldment could be replaced by a single component with reduced weight. Furthermore, the original steel part required surface protection, which was eliminated by switching to a corrosion-resistant non-ferrous alloy. The cost reduction results are as follows:

Current:	Proposed:
Components - 5 pcs. weldment	Components - 1 pc. casting
Material - steel	Material - non-ferrous alloy
Corrosion-resistant surface protection	Surface Protection - no requirements

Weight reduction - 18-20%

Figure 6 illustrates processing cost as a function of casting complexity. As it shows, conversion from ductile iron material to aluminum enables the user to reduce weight up to 45 percent while maintaining the critical characteristics. Moreover, aluminum provides better machinability over ductile iron. Material substitution remains the important cost-saving factor. In a casting process, the material is melted and poured into a mold. Currently, there is a range of process simulation programs which help to achieve desired geometric shape and quality parameters such as good surface finish. Casting offers design flexibility not available in forging or welding processes. In the past decade, casting has been improved by better mold temperature control, multiple-cavity molds, and automation decreased labor. Cost reduction also can be achieved by eliminating machining. When we have reliable casting, dimension control, and fine surface finish, machining can be eliminated.

Today, with applications of advanced technologies, product engineers and designers are able to develop products with better parameters and features at high quality and minimum cost. However, in order to be able to optimize manufacturing processes, they need to be aware of new technological developments. We hope that this chapter will

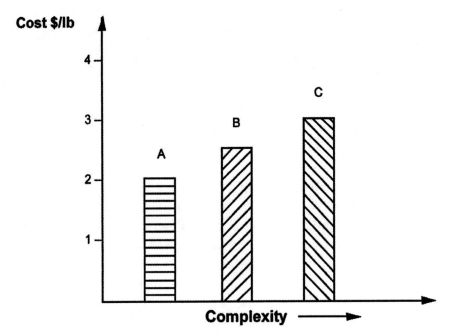

Legend: Weight A = Weight B = Weight C
                    Complexity: A < B < C

Figure 6. Complexity impact on processing cost (casting)

make a substantial contribution to the engineering professional's growth and success.

Recommended sources

1. *Roll forming Handbook* by G. T. Halmos, 2005, Marcel Dekker.

   The book is dedicated to roll forming operations. Process characteristics and machines are clearly described. It is a good source for manufacturing professionals related to roll forming technology.

2. *Manufacturing Applications of Electrochemical Machining: Analysis of Information Flows and Forecasting* by Michael Lembersky. "Electrical methods of metalworking." 1989, p.87.

   The paper analyzes trends in manufacturing applications of electrochemical machining.

   There are some graphical models that illustrate the analytical study results. Electrochemical technological developments are also examined.

3. *Plastic Product Material and Process Selection Handbook* by D.V. Rosato, 2004, 384 pp.

   This book offers a wide range of plastics materials with properties, operations,

   machines used for plastics, molds, and secondary equipment. This is a one-stop

   source for plastic manufacturing professionals for almost all plastic processes.

4. *Advanced Machining Processes* by Hassan El-Hofy, 2005, 253 pp.

   This is a great source of nontraditional manufacturing technologies and applications, such as water jet machining, electrochemical machining, electrical discharge machining, etc.

It contains new processes description and analysis with tables, diagrams, and illustrations.

5. *Process Selection from Design to Manufacture* by K.G. Swift and J.D. Booker, 1997, 214 pp.

   The book presents a wide range of processes and their technical and economic characteristics.

   This is a good source for cost studies and manufacturing selection process.

6. "Cutting Laser Guided by Waterjet." *Manufacturing Engineer,* January, 2005, p. 43.

   This paper describes waterjet processes and laser application for operation improvements.

7. *Computer Applications in Near-Net-Shape Operations* by Andrew Y.C. Nee, K. Soh, Yun G. Wang, 1999. 321 pp.

   This is a good source for completely understanding near-net-shape technology. It contains descriptions of a wide variety of CAD/CAE/CAM applications for sheet metal forming, progressive die designs, injection molding, and related processes. In addition, the book covers CNC system for near-net-shape technological processes.

Recommended Periodicals

1. *Manufacturing Engineering*

2. *Engineered Casting Solutions*

3. *Die Casting Engineer*

4. *Modern Casting*

5. *Metal Producing and Processing*

CHAPTER 2

Knowledge-based system: Technical and economical aspects

Today's products have become so complex that most require a team of professionals from diverse areas of expertise to develop an idea into a machine. Several books are available in the area of materials processing. Most of them discuss the significant practical considerations relevant to the variety of technologies without detailed information underlying economical aspects. A few books (1-2), however, appeared and directed at the physical and chemical mechanism in manufacturing methods. Other books deal with knowledge-based expert systems (3), or design for manufacturability (4). The specific example of a knowledge database is a hydroforming technology Internet database (5). This paper examines the process intended to develop a technological type of knowledge base. In addition, there are several review articles that have appeared particularly on cost estimation aspects, along with methodology.

Engineers, cost estimators, and managers seek to expand knowledge. In order to generate more ideas and turn them into effective manufacturing solutions, we need to expand our technological knowledge. Knowledge-based systems are proving to be a powerful tool with a great potential for developing the intelligent process and material selection system. They allow the user to improve product quality and reduce costs by eliminating or minimizing many of trial-and-error iterations involved in cost assessment. The system integrates knowledge about fundamental technologies and manufacturing operations. It provides powerful reasoning and decision-making capabilities for reducing the time of cost modeling, and consists of the following segments:

- Technological structure of manufacturing processes (principles and operations)

- Data related to machine tools, and presses – descriptions, parameters, capabilities

- Tooling (fixtures, molds, dies)

- Typical manufacturing processes for products and components materials (metals, alloys, plastics, etc.) information.

In light of that, what can a knowledge system (KS) do in support of cost reduction activities?

KS can be viewed as an arrangement of processes, properly put together to transform technical and cost information into finished product cost. It is hard to visualize KS that deals with only a cost development function. Clearly, KS is involved in other business-related costing activities, such as:

- Design concept evaluation

- Engineering change cost control

- Supplier evaluation

- Competitive teardown analysis

- Cost requests

- Cost reduction analysis

- Alternative cost studies

- Features and options cost assessment

- Value analysis

- Target cost setting

- Cost estimation of outsourcing

- Design for manufacturability – costing support

- Cost advice or cost recommendations

Switching from one product to another requires cost justification that can be accomplished efficiently only based on knowledge systems (KS).

It should be noted that the scope of cost estimator responsibilities is not limited by daily activities. It includes the development of policies and forms which will speed up the cost response. Moreover, management recently pays significantly more attention to having the technology center. This center contains data about best practices, advanced processes, and equipment characteristics, magazines and trade journals with updated engineering and pricing information, and process comparison information. It is also important to have a segment related to cost estimation software. Recent developments in computer modeling created a number of cost software programs. Typically, there are two general types of software for cost estimating:

- Integrated software packages containing a set manufacturing processes, materials, tooling cost

- Process-related estimation package for stamping, casting, machining, etc.

The integrated software package allows the user to carry out cost estimation of a product which consists of plastics and metal components that have been manufactured by injection molding, extrusion, forging, and many other methods. All those methods, however, are included in one cost estimating software package. Process-related cost software includes only individual manufacturing processes, for instance, only stamping or casting. When integrated software is applied, processing cost evaluation begins with process planning that entails the specification of the sequence of operations required for converting raw material into parts and then assembling parts into products. Process selection requires an in-depth understanding of the availability and limitations of different manufacturing processes. One of the questions is how material will flow through the system and how processes are linked to obtain the desired volume of production at the intended quality. To find an efficient material forming solution to customer product, a practitioner analyzes the process categories available for the specified material at certain surface finish and tolerance levels. There are several hundred individual manufacturing technologies commercially used in the production environment. Therefore, the results from a systematic approach to the best process selection with engineering vision can be very beneficial in terms of economical deliverables. Figure 7 illustrates relationships between creativity and engineering science, which are usually applied to present the engineering view of a product. In addition to that, it enables the user to visualize the entire set of the data included in a comprehensive alternative system analysis and their interactions. That is one of the concepts that represents knowledge system abilities to direct a manufacturing task with a systematic approach. A practitioner synthesizes all factors related to the certain task. When the software does not have a built-in tool to keep track of the entire costed BOM, Excel workbook allows the user to properly present or generate the standard BOM devoted to a product line. In order to cost certain features, a cost estimator or assigned engineering professional can add or delete components based on modified engineering BOM. Again, the complete information about materials and process parameters, machines, and tools can be found in a knowledge system that includes a technological subsystem and product subsystem. Figure 8 illustrates the visual model of a documentation subsystem. This model clearly demonstrates all components that are included in a complete set of costing documents. When a company is developing such a comprehensive system, it provides teams and departments with accurate data allowing them to generate product target cost, put together alternative cost scenarios, "should cost" models, and keep consistency across all costing activities. The basic groups of manufacturing processes (Figure 9) need to be presented in the system. In addition to that, there is a typology of materials (Figure 10) used to manufacture most of the products and components. The systematic approach with engineering vision enables the user to analyze and select the optimized operation sequence at minimum cost. The knowledge system, which includes both material properties and processes, should be constantly maintained and upgraded with new information related to machines, robots, fixtures, and their characteristics and costing information. The knowledge system of processes can be also presented as follows:

- Material removal processes

- Forming processes

- Phase-related processes

- Structure-related processes

- Secondary processes

- Finishing processes

- Joining processes

- Cross process combinations

Under material removal process, we mean manufacturing processes, such as turning, milling, grinding, boring, drilling, etc. Currently, there are machining centers with capabilities of operational concentration as follows: milling and turning of multiple surfaces with high precision requirements. The benefits are short cycle time and high productivity. Another type of machine with high concentration of operations is a rotary transfer machine, which specializes in high-volume production of irregular-shaped components machined from castings, forgings, or blanks. This type of machine can perform up to thirty machining operations simultaneously, on up to three axes at each station. The capabilities of accommodating part redesign, manufacturing of complex parts, and short cycle time indicate high potential in terms of cost reduction opportunities.

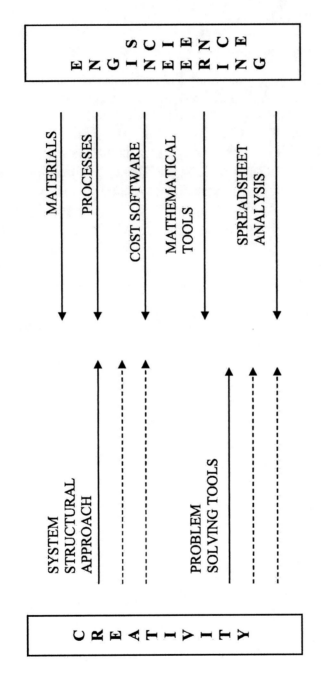

Fig. 7. Creativity and engineering science relationships.

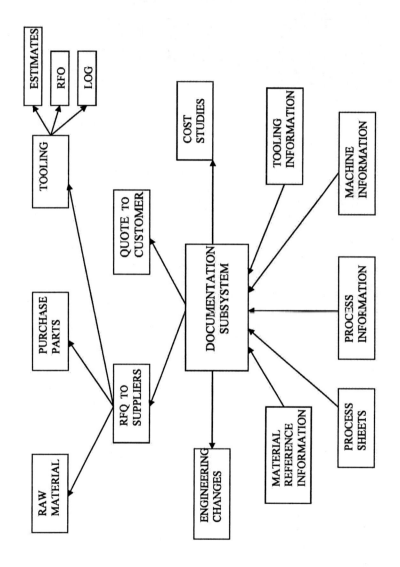

Fig. 8. Documentation subsystem visual model.

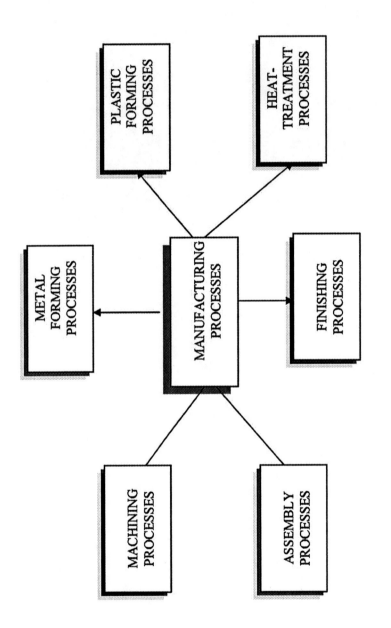

Fig. 9. Groups of manufacturing processes.

46

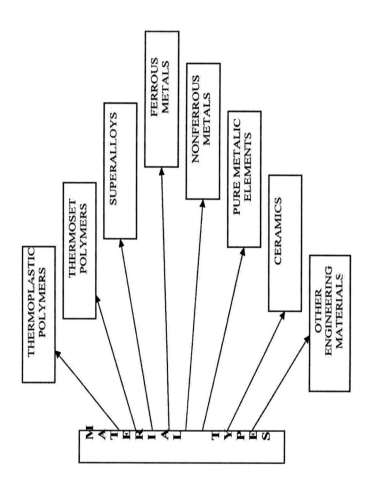

Fig. 10. Typology of materials.

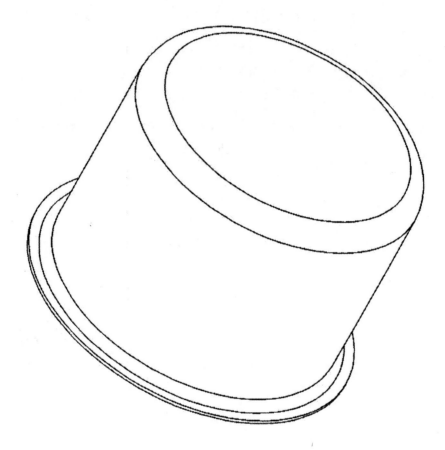

Fig. 11. Part manufactured by spinning process.

The metalforming processes are applicable when the shape of the part needs to be changed without changing its mass or composition. The type of part manufactured by the spinning process is shown in Figure 11. Forging, sheet-metal forming, shearing, drawing, extrusion, and rolling are representatives of this category. In the past decade, there have been significant improvements in press characteristics, dies, and computer applications for better process control and accuracy, as well as increased productivity. All these factors enabled the user to raise the competitiveness of metalforming technology among other manufacturing processes.

Material phase-related processes are casting processes, metal injection molding, plastic injection molding. In other words this category of processes are intended to make a solid part from liquid phase of material. The casting process and equipment improvements allowed to replace multiple components assemblies with one or two parts: one casting instead of several stamped and machined parts (see Appendix to this chapter). The plastic injection molding (PIM) helps to reduce cost by replacing metal parts with plastic components. PIM is also applied as a simultaneous process, when two normally separate steps are combined in one shot operation.

The benefits are: improved dimensional integrity, quality, reduced cycle time, tooling cost and scrap. The computerized control systems of plastic molding machines also enables the user to improve the product quality and process efficiency.

Secondary processes include: deburring, trimming, surface preparation (cleaning, degreasing). This category of manufacturing processes was developed in two directions, such as application of new physical and chemical principles and computer control systems. Now there are many more companies that are applying electrochemical, thermal, and abrasive flow deburring because of the increased productivity and reliability of these processes. A structure-related category of processes is applied when it is necessary to change part microstructure or hardness. In other words, heat treatment types of processes, which at the present time are equipped with computerized control systems, enabled the user to achieve a wide range of surface characteristics. The category of finishing processes represents:

plating, painting, coating, and so on. These processes are capable of accomplishing a very high level of surface quality with high wear resistance using advanced automation systems and machines.

The category of joining processes include welding, fastening, and a variety of assembly processes. A wide range of commercial automated applications in welding technology were developed and implemented which enabled the user to use advanced materials with new properties. The category of cross process combinations include electrical discharge machining, electrochemical machining, water jet machining, laser cutting, plasma cutting, and so on. In the past decade, there were many new processes based on combinations of traditional and nontraditional processes: electrochemical grinding, electrochemical honing, abrasive jet machining, and others. Application of combined processes enables the user to design better products from advanced materials with more features and functions.

The systematic approach to manufacturing cost includes material analysis. The entire material system can be classified as it shown in Fig. 10 Many years ago, the material cost could be determined easier than today, when we have enormous material types and grades and prices. In addition to that, the price is highly sensitive to the grade and annual production volume, for example:

Plastic material price = $1.35/lb.

Higher grade material = $1.46/lb.

Annual material usage = 500,000 lbs.

The cost variance: ($1.46/lb - $1.35/lb) * 500,000 lbs. = $55,000.

This calculation indicates how material change applied to a component helps to save on material cost. In other words, if it would be possible to replace material with less expensive material, cost savings would be $55,000. The knowledge of material systems and properties enables the user to effectively conduct material selection process, to assist in material substitution for cost savings purposes. The importance of material knowledge is greater with multiple-material products as well. In order to advise on material cost decisions, cost responsible specialists need to know where to find information about a specific material, at which shape this particular material is available. All these data associated with material help to make appropriate trade-offs between design requirements and material cost on the market.

The important role in optimizing the planning process for manufacturing cost study and analysis is the use of empirical knowledge. In order to link empirical data in an appropriate way, the knowledge system is needed. The empirical data can exist in forms and logs, drawings, worksheets, and published sources. Such data can be organized and entered into the system. After that, information can be searched by certain attributes (for example machine types, parameters, material types, properties, geometric characteristics, raw material shapes and profiles, process types, descriptions and empirical formulas for cycle time). The knowledge system can be divided into several major interconnected modules:

- Machines database

- Material database

- Manufacturing processes database

- Sales database

- Marketing database

- Best practices database (particularly related to product-specific technologies)

- Literature database (references, periodicals, books, etc.)

Today, many companies develop Web-based sources. This structure can be suitable to Web sites as well. We need to keep in mind that the prospective benefits of having a well-organized knowledge system are:

- Effective material and process selection

- Support of engineering change process

- Continuous product improvement process and efforts
  make or buy decision support

The knowledge system allows the user to conduct an in-depth process analysis in terms of its capabilities to achieve required component characteristics, productivity, and reliability. The system also supports selection of the most effective process and the most economical decision in new product introduction, competitive analysis, sourcing solution, or engineering change process. After applying the knowledge system, a practitioner is able to discover new ways to think about existing processes and generate new process configuration.

Recommended sources

1. *Die Casting Engineering: Hydraulic, Thermal and Mechanical Process* by Bill Andersen, 2004, 384 pp.

   This book covers principles, applications, and machines for casting technology. A clear style makes the book very useful for practitioners.

2. *High Integrity Die Casting Processes* by Edward J. Vinarcik, 2003, 223 pp.

   The book provides a comprehensive look at the concepts behind advanced die casting technologies. Practical applications of squeeze casting and vacuum casting are also presented in several case studies.

3. *Manufacturing Database Management and Knowledge Based Expert Systems* by Paul G. Ranky, 1990, 237 pp.

4. *Design for Manufacturability* by Dr. David M. Anderson, 1990, 225 pp.

5. "Building a Web-based Database of Hydroforming Knowledge." *Steel Times International*, 10/2002.

6. *Engineering Management* by Fraidoon Mazda, 1998, 658 pp.

   The book provides very important information about strategy and decision-making and operations management for engineering professionals and managers. There are many examples, tables, and mathematical formulas and illustrations.

CHAPTER 3

Cost estimation process: systematic approach

The lion's portion of business success depends on the product cost. The estimation comes into play for several reasons: to prepare budgets, for cost management, price calculations, and to make offers to customers. Revenue is needed to cover the cost and to generate profit. Winning the profit game is hard. Evaluation of product performance is based on cost system; therefore, it must be aligned with financial system and manufacturing capabilities. This is required to allocate resources optimally, set priorities, and achieve advantage over competitors. If enough resources are dedicated to develop, analyze, and control the product cost, a company will be able to select the best business strategy and move in the right direction. Another statement is true as well: Goals can be achieved when the effective cost system is in place. Existing cost modeling methodologies vary in scope, appearance, and theoretical foundation. COST ESTIMATION has been always called an art and a science. The purpose of this book is to bridge art and science, which can be achieved by mastering science and supporting an art side of cost estimation. Cost estimation provides answers to many questions: whether to apply the same manufacturing strategy to a new product or not, choose whether to make a product in house or outsource, define your competitor's cost position, etc. Those answers enable the user to predict competitiveness and develop cost reduction strategy.

THE COST ESTIMATION PROCESS is dependent on many factors. Only the systematic approach to the entire costing process enables the user to come up with the optimal solutions that will serve the business's goals. In order to manage a costing system properly, a company analyzes cost methods and models in terms of their actual connections to the particular business model. The importance of an effective cost system is to provide accurate, relevant deliverables in support of business, as well as to help with problem solutions. That means the system must contain tools and resources to accomplish its goals. Presently, most businesses are generating costs based on spreadsheets and software programs at a high level of detail, and compiled in the workbook. Usually a print, product specifications, samples, reference information and spreadsheets are among the main components of any cost studies. Recently, great progress has been made in providing businesses with new methods for costing process.

There are many different types of software packages developed to automate the cost estimation process. Some of the software programs contain information on machines and tools for the wide variety of manufacturing processes, while other programs contain data on the specific process and its modifications. In other words, presently, cost software products have a wider functionality than before. So, in order to make a better business choice, the user analyzes available costing software from a broader perspective. He creates a list of preferred key parameters that enables him to choose the right options.

Why are companies focused more than ever on the streamlining of estimation procedures? Because the increased complexity of business functions demands timely cost information for the optimal decision.

Let us start from the beginning. The product cost structure can be presented as follows:

- Material cost (raw materials, purchased components, outside services)

- Direct labor cost

- Overhead cost

- General expenses

There are products where the material cost portion is 45-55 percent of the total cost. This indicates that accuracy of material cost calculations is critical. Whether it's a print or a sample to be estimated, identification of material type is the first step. There is a significant cost difference between steel, plastics grades, magnesium, and aluminum. Magnesium is an engineering material with the lowest density. Despite its high cost and limited supply, industrial applications are rapidly growing. On a per-pound basis, magnesium about two times more than aluminum, but on a volume basis, it costs only 1.2 times more. It should be noted that cost comparison of competing materials is often misleading because of the difference in materials' performance and density:

Magnesium 1.8 g/cm^3

Aluminum 2.7 g/cm^3

These data indicate that magnesium's density is about two-thirds that of aluminum. Magnesium has a lower Young's modulus than aluminum, which will require using more mass to obtain the same rigidity (safety margin factor to ultimate tensile strength) in the magnesium part. When we compare materials costs, it is substantial to compare materials properties as well. The multifactor analysis will allow the user to optimize the material selection process.

Currently, material cost rates can be found on Web sites concerned with plastics, steel, or cast iron. The material specifications, however, are required in order to be able to use the Web site data. You can find steel costs, if you know exactly the type of steel, shape, and mechanical properties (strength). The information received in such details really helps to come up with correct material cost. There is one more important feature: When switching from one material profile to another, you are able to instantly determine which material type enables the user to reduce piece cost and increase quality. The query research is based on material characteristics. There are Web sites which are helpful in finding various materials based on given strength, hardness, or other characteristics specified by print. These are some advantages of using Web sites for material selection in alternative cost studies or any cost-related projects.

Purchased components include parts, subassemblies, fasteners, and electronic devices. Prior to sourcing decisions, it is important to conduct a supplier search in terms of better price, reliability, and quality. In addition to that, factors concerned with supply chain management, logistics, and packaging require in-depth considerations, because oftentimes, they play significant role in final cost. The optimal configuration of the supply chain allows the user to save money. When a manufacturer receives from three suppliers a bracket, a steel cover, and a subassembly, it needs to consider the optimal equation in order to minimize purchase cost. If both bracket and steel cover are stamping parts as well as the subassembly, it can be more effective to find one source to manufacture the entire set. This decision is going to reduce packaging cost and freight cost, and improve quality. The optimization analysis of the purchase portion of cost improves the supply chain and reduces cost accordingly.

Purchasing cost also includes cost of services which are required for manufacturing of product, but the company is not specialized in these particular operations; for example, heat treatment is specified by a customer for a component. The company is equipped with stamping presses, but not with furnaces, so heat treatment is provided by an outside supplier. The cost affected by subcontracting this operation should be minimized by a search for potential suppliers capable of providing the required type of heat treatment with lower cost and higher quality.

Clearly, purchase cost contributes substantially to product cost. That is why a cost justification study is required before placing the order with a supplier. Direct labor cost contains operator cost directly related to manufacturing a component or product. The following equation is representing labor cost calculation in general view:

$L = E \times (R \times T)$,

Where:

> L – labor cost
>
> R – labor rate (labor cost per unit of time)
>
> T – time units used
>
> E – composite factor (allowances)

Labor cost rates are based on manufacturing type of operation and skill level. The operation time depends on pure machine time and allowances, which enable the user to analyze the effectiveness of manufacturing. In the real world of manufacturing, there are a number of work interruptions, so labor allowances must be added to calculate labor cost. When a component goes through several operations, the total labor cost can be calculated as follows:

$L = L1 + L2 + L3 + \ldots + Ln$,

Where:

> $L1, L2, L3, Ln$ = labor cost on each manufacturing operation (forging, machining, coating, etc.).

Labor cost is based on the analysis of the entire sequence of operations and steps associated with product manufacturing. The industrial engineering department develops labor standards and conducts time studies, which can be obtained by a cost estimator in order to generate accurate cost.

Burden cost is a group of cost elements that is a very important contributor to product cost. In order to calculate the machine cost portion, we need to multiply machine rate by time that certain machine or press is used to make a part. General considerations for a piece of equipment (machine) are power, size, and capacity. The information for machine rate calculation includes the following:

- Machine specifications

- Machine floor space

- Number of parts per cycle

- Total part lifetime volume

- Annual working hours

- Machine uptime

Machine parameter or press tonnage, for example, in sheet metal (steel) stamping operations for preliminary purposes can be determined based on the formula:

Material Thickness x Inches of Cut x 25 = Press Tonnage

where factor 25 is shear strength (25 tons/square inch for steel).

Machine rate includes portions related to machine purchasing cost, installation, machine technical time, depreciation, and residual value. Utilities (water, gas, electricity) are also part of overhead cost:

utilities cost, $ = rate x usage

Based on this formula, we can calculate the cost of water, gas, and electricity separately. The cost of machine floor space, insurance, machine maintenance, repair, and cleaning have to be added to calculate overhead cost as well. In spite of so many elements of overhead cost, all of them have to be taken into consideration for the purpose of cost accuracy. Stamping press cost, for example, can be calculated as follows:

Annual parts volume 150,000 pcs.   Description: 500-ton press

Run rate 1,500 pcs./hr.   Machine rate: $114/hr.

Calculations:

> Conversion to pcs./min.
> 1,500 pcs./hr./60 min. = 25 pcs./min.
>
> Conversion to machine rate per min.
> Machine cost = $1.90 x 25 pcs./min.

The following example represents cost estimate with typical sequential steps involved:

> Material usage can be determined by using drawing, specifications, or a sample.
>
> The material cost = material rate x part weight = R * M
>
> > Where material rate R is the cost per unit of measure ($/lb., $/kg, $/ft., $/in.), the part weight M is expressed in lbs. or kg.
>
> The part weight can be calculated based on volume and material density.
>
> The part weight = material density x volume = D * V
>
> > Where D is the material density (lbs./in.^{3}, g/cm^3), V is the volume (in.^{3}, cm^3).

However, when you deal, for example, with a piece of tubing, the rate can be expressed in $/ft., $/in., or $/m because the material usage is expressed in linear dimensions, which are feet, inches, or meters. Thus, the formula for the material cost is:

> Material cost, $ = material rate $/ft. x linear dimension, which is the length expressed in feet.

Oftentimes, the shape of a part or component is a combination of a cylinder, cube, or other geometrical shape.

The volume of such component can be expressed as follows:

$V = Va + Vb$

Where Va = volume of cylinder, Vb = volume of cube.

In a situation where the shape of component represents two cylinders, use the formula:

$V = 2 \times Va + Vb$

The drawing or specification contains most of the data needed to develop the cost estimate.

A designer applies variety of different materials (steel, aluminum, zinc, plastics, composites, cast iron, etc.). It is very important to follow the specific material description, standard (SAE, ASTM, etc.), because weights and material rates depend on specific material properties. There are two main types of materials in respect to manufacturing and production: direct and indirect materials. Direct materials include raw materials purchased and not manufactured by the company. The manufacturing process requires welding supplies (rods), lubricants, grease, etc., which are indirect materials. Indirect materials are not a part of the final product. The basic information about raw materials includes description – steel, cast iron, aluminum, or magnesium, etc. What type of steel – low carbon, stainless, galvanized, etc. There are different standards that are used by designers in order to specify raw material (SAE – Society of Automotive Engineers, ASTM –American Society for Testing and Materials).

Even if you work for a company that deals only with a narrow range of materials, you would still need basic information about other materials in case of an alternative cost study. So the knowledge of materials standards is important in many aspects. However, building your own source of specific information or a database will help you to avoid some very time-consuming steps.

We would like to provide you with a hypothetical example of cost estimating. The cost equation can be presented in the following view:

PrC = MC + OC + LC + BC

> Where PrC = product cost, MC = material cost, OC = outsourcing cost, LC = direct labor cost, and BC = burden cost.

As it was indicated, material cost can be calculated as follows:

Material: SAE 1008-1010 steel

Material cost: MC= R * W

Material rate: R = $0.31/lb

Part weight: W = 0.9 lbs

Calculations:

MC = $0.31/lb. * 0.9 lbs. = $0.28

The specific approach to the cost model development depends on a type of the process. In other words, in the case of stamping or forging, the cost will depend on process-related cycle time per piece. The following example presents the calculation of labor cost and burden cost:

Every minute, 80 parts come off the press.

Press hourly burden rate - $75/hr.

Operator rate - $25/hr.

Calculations:

Cycle time = time/parts   60 min./80 parts = 0.75 min./
                                           piece

Labor cost = Labor rate x Cycle time
Conversion from cost per $25/60 min. = $0.42/min.
    hour to cost per min.
LC = $0.42/min.*0.75 min.*1 = $0.32

Burden cost = Burden rate x $75/60 min. = $1.25/min.
    Cycle time
Conversion from cost per   BC = $1.25/min. * 0.75 min.
    hour to cost per min.   = $0.94

The next step is to calculate outsourcing cost. What do we really mean when we are talking about outsourcing cost? Every single part or component is subjected to variety of different operations in order to be shipped to a customer as a finished product. A bracket, for example, according to print requires a certain type of coating that can be done only by a specialized company. A potential supplier quoted this bracket for $0.25/pc. So the coating cost represents the outsourcing cost, which is going to be included in the product cost estimate.

Finally the total cost of the bracket:

PrC = MC + PC + LC + BC = $0.28 + $0.25 + $0.32 +
    $0.94 = $1.79

In reality, there are many different types of products, when every single cost element requires an individual approach. To reinforce everything that was written above, we present an example; when a product is an assembly of several components, the product cost is going to be a sum of individual component costs that go into this assembly. For example, assembly consists of four parts: two aluminum parts attached to the plate. The goal is to define the product cost. The next step would be to develop the product cost structure of this assembly:

PrC = $P1 + $P2 + $P3 + ASSY COST

The product cost of the plate assembly consists of the cost of each component and assembly operation cost.

Previously, we walked you through the cost calculations related to a single component.

Assume that the cost of parts included in this assembly:

Part 1: Material – Aluminum

Weight - 0.35 lbs.
Rate - $0.75/lb.
Material cost: 0.35 * 0.75 = $0.26
Cycle time: 9 sec. / 60 sec.= 0.15 min.
Labor rate - $25/hr. / 60 min.=$0.42/min.
Labor cost: $0.42/min. * 0.15 = $0.06
Burden rate: $70/hr. / 60 min. = $1.17 min
Burden cost: $1.17 * 0.15 = $0.18

Total manufacturing cost:
$0.26 + $0.06 + $0.18 = $0.50/pc

Part 2: Material – Aluminum

Weight - 0.28 lbs.
Rate - $0.75/lb.
Material cost: 0.28 * 0.75 = $0.21
Cycle time: 8 sec. / 60 sec.= 0.13 min.
Labor rate - $25/hr. / 60 min. = $0.42/min.
Labor cost: $0.42 * 0.13 = $0.05
Burden rate: $70/hr. / 60 min. = $1.17 min.
Burden cost: $1.17 * 0.13 = $0.15

Total manufacturing cost:
$0.21 + $0.05 + $0.15 = $0.41/pc.

Holder – Material: Nylon

    Weight – 0.9 lbs

    Rate   - $0.75/lb.

    Material cost: $0.75/lb. * 0.9 lbs. = $0.68

    Cycle time – 11sec./60 = 0.18 min.

    Labor rate $25/hr. * 60 = $0.42

    Labor cost: $0.42 * 0.18 = $0.08

    Burden rate: $80/hr. /60 = $1.33/min.

    Burden cost: $1.33 * 0.18 = $0.24

Total manufacturing cost: $0.68 + $0.08 + $0.24 = $1.00

Assembly cost of all three components and four fasteners:
    Fasteners cost: 4 pcs. * $0.04/pc. = $0.16

Assembly operation time: 12 sec. /60 = 0.2 min.

Labor rate: $ 25/hr. / 60 = $0.42/min.

Labor cost: $0.42 min. * 0.2 min = $0.08

Burden rate: $ 30/hr. / 60 = $0.50/min.

Burden cost: $0.50/min. * 0.2 min = $0.10

Total holder assembly cost: $0.41 + $0.50 + $0.16 + $0.08 + $0.10
    = $1.25

COST COMPARISON BREAKDOWN

Company: _____
Date: _____
Program #: _____

Product Model: _____
Part #: _____
Revision #: _____

Annual Volume: _____
Part Description: _____
Capacity: _____

Proposal 1

Description	U of M	Qty	Extended Cost
RAW MATERIAL			
Steel			
Polyethylene PE			
Sub Total			
PURCHASED ITEMS			
Bolt M10x1.5			
Nut			
Washer			
Sub Total			
LABOR & BURDEN			
Injection Molding			
Assembly			
Sub Total			
MFG TOTAL COST			
END ITEM SCRAP			
G.S. & A.			
PROFIT			
R.D. & E.			
FREIGHT			
OTHER			
GRAND TOTAL COST			

Proposal 2

Description	U of M	Qty	Extended Cost	Variance
RAW MATERIAL				
Aluminum				
Polypropylene PP				
Sub Total				
PURCHASED ITEMS				
Bolt M10x1.5				
Nut				
Washer				
Sub Total				
LABOR & BURDEN				
Injection Molding				
Assembly				
Sub Total				
MFG TOTAL COST				
END ITEM SCRAP				
G.S. & A.				
PROFIT				
R.D. & E.				
FREIGHT				
OTHER				
GRAND TOTAL COST				

Fig. 12. Cost comparison breakdown (hypothetical example).

This example represents the entire process of product cost estimation. When the same format is applied in different cost requests or studies, then consistency of the cost process improves business planning and the decision process. One of the hypothetical examples of cost comparison is presented in Figure 12.

It is hard to overemphasize the importance of methodology in the modern costing process. There are a wide variety of business activities that require costing support in terms of studies, spreadsheet analysis, etc. If these studies are based on best practices methodology, then business activities will flow in the right directions and will be successful and profitable. The improved cost methodology helps to set priorities and find ways to reduce cost and follow targets accordingly. Furthermore, practitioners need to be aware of advanced processes and their capabilities in terms of manufacturing operations, productivity, and other factors. Knowledge of machine technical characteristics allows the user to calculate and analyze overhead expenses associated with manufacturing. Generally, overhead expenses include:

- Machine cost

- Maintenance cost

- Building cost, insurance cost

- Depreciation cost

- Utilities cost

Machine cost represents the cost of machine floor space, machine specifications, and number of parts per cycle. The remaining overhead elements can be calculated based on data from product cost system, where information in terms of rates and usage is updated and tested.

The general view of the production system is structured as follows:

- The workpiece

- Manufacturing process (equipment, tooling, operator)

- Final product (assembly or component)

In order to succeed, management organizes the most effective cost estimating system to be able to control cost, to optimize resources, and to lead the stable company growth based on a reliable cost system. What cost modeling means is relating process parameters to process cost. The development of the product cost begins with the process planning:

- Analyze cost assumptions

- Review the entire set of documents and specifications

- Develop BOM (bill of materials)

- Record questions you have to customer

- Develop work breakdown structure based on provided information

- Formulate cost estimating relationships

- Search for reference information

- Develop Excel spreadsheet or use a company template

- Use in-house developed cost-estimating software

- Complete the estimate

Steps will vary from one estimate to another, which is dependent on the type of request: customer RFQ (request for quotation), supplier quote review, marketing or sales account manager wanting to conduct alternative cost study, design engineer needing to know the difference in costs between design concepts. The engineering and purchasing departments oftentimes are looking for an estimate related to engineering change. The customer is going to make product changes. Therefore, steps for estimate development will vary from study to study. However, main steps cannot be missed, otherwise the cost is not going to reflect all resources it takes to manufacture a product.

At the same time, every cost estimate requires an individual approach. Let us assume a practitioner is assigned to conduct a study. He takes into account that the use of press or a machine tool will require a different skill level and number of operators, and this is going to affect labor cost. For example, applying a machining center for milling and turning a part reduces labor cost compared to making a part by using a milling machine and lathe in sequence. In the case of tooling operations, when an electrical discharge machine is applied to create a cavity in a mold, you can save labor cost as well, because an EDM machine is able to replace several traditional machine tools with just one piece of machinery. When it comes to evaluation of benefits, a company can achieve by implementing advanced technological innovation, but it really requires in-depth analytical investigation for the most effective business solution. The optimized solution is based on analysis of factors from technical and economic aspects.

The important part of the entire cost analysis is tooling cost (Appendix). Tooling cost development is based on generating a workbook. The main cost elements of tooling cost are:

- Design cost

- Material cost

- Build cost

- Purchased components cost

- Outside cost

- Tryout cost

Sometimes, for preliminary die cost estimation, practitioners are using parametric cost; an example could be the cost of one station in case of progressive die. One station may cost $6,000. So if a progressive die consists of seven stations, the cost would be $42,000. However, for more accurate tooling cost, including dies, the above cost structure is needed.

The cost practitioner is involved in cost details and searching for data to support calculations. When you have a detailed work breakdown, you are able to deeply analyze the cost associated with individual elements such as material or labor. Furthermore, when a customer initiates an engineering change, you are able to pull out the original estimate and conduct the study based on previous calculations and assumptions. When it comes to analysis, it is extremely important to find out about all cost assumptions. In other words, when the costing process is organized, well-documented, and consistent, including assumptions, it helps to speed up the engineering change costing process, while at the same time, to build reliable relationships with the customer.

The practitioner also performs reverse engineering when developing a cost model. He conducts manufacturing type of analysis in terms of operations and machines applied to convert workpiece to finished component. In the case of competitive study, the practitioner compares cost to cost, feature to feature, function to function, performance to performance.

CHAPTER 4

Cost reduction process: methodology and analysis

Cost reduction efforts are very important for enhancing profitability and cash flow. To date, most work carried out in the manufacturing cost analysis has focused on identification of cost reduction opportunities. Although many practitioners have different official areas of specialization, they oftentimes are connected to the cost assessment process. They need to gain more control over product cost reduction (CR) when introducing new products. Cost reduction and process improvements can target a wide spectrum of possibilities, including supplier specifications and process sequencing. In addition to that, CR is a multivariable process (Fig. 13). Formulating CR strategy has to be generated with engineering vision. In recent years, the use of cost methodology as a base for CR has been significantly increased. At the same time, influence of detailed cost evaluation is becoming more and more critical to the decision-making process. Based on the knowledge of principles and concepts of developing product cost, a practitioner is able not just to put together a cost estimate, but also to carry out the product cost analysis and determine the major cost drivers for the purpose of cost savings. What are the reasons behind the cost modeling? Why do we need to know the cost?

- First and foremost, cost is a measure of resource consumption.

- Cost, as an objective measure, is useful for accounting, measuring revenue/profit.

- Engineering needs a cost tool to evaluate current process conditions or new alternative concepts.

- Business planners need to evaluate investment needs and opportunities.

- Management needs to evaluate competitiveness of the product, to make the best economic decision.

At the same time, the cost estimate is the key structural component in the continued growth of the company, because without an accurate economic view, the company cannot go forward. The accuracy and effectiveness of any cost estimate depends on:

- The time dedicated to prepare the cost model

- The knowledge and skill of the estimator

- The depth and completeness of the collected information

- The knowledge of the best practices.

As mentioned earlier, there are always three types of players (Fig. 14) who interface with each other: supplier, manufacturer and customer. The cost assessment process helps each of them to effectively control resources, as well as to save money and to keep good relationships. A systematic approach to product cost reduction analysis begins with the scope of the proposal.

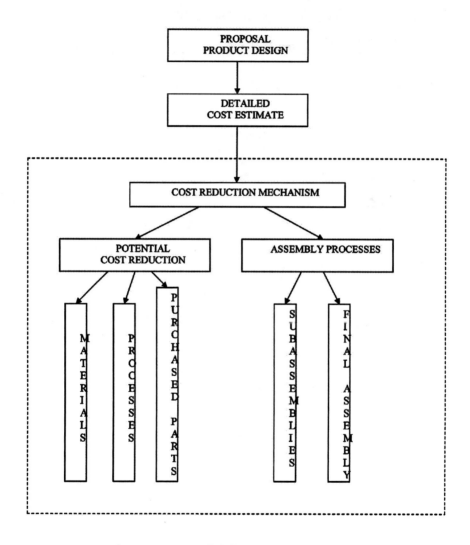

Fig. 13. Cost reduction: conceptual view

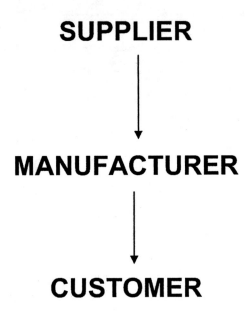

Fig. 14. Main players

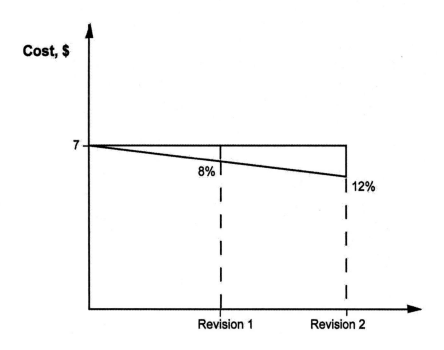

Fig. 15. Revisions for cost reduction

For example, the product team's task is to conduct a cost reduction study. Figure 15 shows a hypothetical product revision that enables the user to reduce the product cost by 8 and 12 percent. This visual representation helps to emphasize the importance of proposed cost-saving ideas. Visual modeling became a very important methodological tool for analysis and communication purposes in cost reduction and training purposes.

Let us say a full-blown cost estimate has been generated in a way that required detailed calculation of each cost driver, such as: raw materials, purchased components, outside services, processing cost, and overhead cost. A proposed product cost model has been generated at the same level. Then a cost item comparison (Fig.16) has been performed, based on bills of materials of both products: existing and proposed. Major components have been proposed made of plastics with incorporated features instead of fasteners. Therefore, parts have been made using an injection molding process. The cost study results presented to the cross-functional team enabled them to conduct an in-depth value analysis, which led to cost reduction changes. Let us calculate the total savings based on 150,000 units per year: ($1.83 - $1.26) * 150,000 = $85,500. The implementation of all necessary improvements will help a company to save $85,500 on one product. This significant cost reduction opportunity also indicates the importance of well-thought-out cost estimation processes. Another way to reduce cost is to communalize different products. First of all, a team generates a list of potential products in terms of their similar features, materials, and functions. Then, the next step is to select the product that would represent better than others the product line in respect to features, functions, and manufacturability.

There is a variety of ways to carry out the communalization process. However, from the knowledge-based point of view, it requires one to investigate carefully possible directions in terms of engineering perspectives. For instance, analysis of "best manufacturing practices" allows the user to find out about nontraditional process. Electrochemical machining (ECM) enables the user to reduce tooling costs (plastic injection mold) by 10 - 15 percent, because ECM is able to eliminate several machining operations. ECM does not depend on the tool material hardness. It electrochemically dissolves the surface and generates the required shape of the cavity with a high quality level and tight tolerances. Thus, the use of nontraditional manufacturing processes allows the user to reduce cost and improve the product design as well. Because in many cases it enables the user to incorporate a feature that requires additional shape-related complexity in a mold that is not achievable with traditional machining operations.

#	PRODUCT A (Current)			PRODUCT A (Proposed)		
	PART A1			PART A1		
1	ALUMINIUM	$0.51		PLASTIC MATERIAL	$0.43	
	CASTING (PLUS OVERHEAD)	$0.49		INJECTION MOLDING OPERATION	$0.56	
	MACHINING (PLUS OVERHEAD)	$0.43				
2	PART A2			PART A2		
	BRACKET 1	$0.28		BRACKET 1	$0.20	
	BRACKET 2	$0.12		BRACKET 2	$0.07	
	TOTAL	$1.83		TOTAL	$1.26	

Fig. 16. Alternative design cost comparison.

82

All cost estimator's findings are very beneficial to the cost reduction process especially, if they are based on knowledge and expertise in the advanced manufacturing technology. Besides that, there are cost minimization methods such as: use, use of preferred fits, preferred component dimensions, and commonly used standard parts. The structural view of CRM presents the interconnected elements which are the targets of cost reduction. Engagement in cost reduction efforts leads to bridging the gap between creativity and engineering science of estimation. Before analysis begins, the practitioner finds answers to the following questions:

- Will using a new process eliminate or reduce some operations, including secondary operations?

- Will using a new process allow me to use less expensive materials or to stop wasting as much material?

- Will using a new process greatly increase production rate?

- Will using a new process provide higher performance or a higher-quality part?

In the Appendix, you can find, as we mentioned earlier, major materials representation. According to this diagram, the important step in product material analysis is to find to which group each component material belongs. For instance, print specifies a plastic material – nylon. The practitioner, who is in the process of cost analysis, investigates properties and costs of raw material. The material database of PCS enables the user to speed the research phase. Otherwise, data can be found in a reference source or on a Web site. Thus, repeatability and consistency in the estimation process enables the user to create accurate models. There are two types of changes: incremental and innovative.

Incremental change

A component has been redesigned in accordance with new customer specifications. The engineering change enabled the user to apply a smaller press, which was a cost-saving factor. However, forming type of process remained sheet metal stamping. Figure 17 represents a form developed for an incremental change, where an aluminum component has been added to a subassembly, while a steel part and a plastic (nylon) part have been eliminated by design improvement. Clearly, this type of form can be used for a wide variety of quick preliminary cost evaluation.

COST ITEM	ADD	DELETE
MATERIAL		
STEEL		$0.85
ALUMINUM	$0.69	
NYLON		$0.89
PURCHASED PARTS		
BOLTS		$0.16
NUTS		$0.16
TOTAL	0.69	2.06

Fig.17. Incremental change cost form.

Cost Improvement Opportunities	Cost Saving Ideas	Preliminary Estimation		Total Cost
		Quantity	Cost Per Unit	
1. Common Components	1.			
	2.			
	3.			
	4.			
2. Materials Substitution	1.			
	2.			
	3.			
	4.			
3. Part Consolidation	1.			
	2.			
	3.			
	4.			
4. Non-traditional Processes	1.			
	2.			
	3.			
	4.			
5. Other Opportunities	1.			
	2.			
	3.			
	4.			

Fig. 18. Preliminary cost analysis.

The increased complexity of the manufacturing environment is the main reason for changes in cost reduction methodology (CRM). Methods applied to cost reduction include: cost modeling, product functional analysis (PFA), and manufacturing technology evaluation (MTE). Cost modeling is improved by implementing new software packages. Product functional analysis presents the entire picture from value analysis aspects. Value analysis based on engineering analysis, that requires application of MTE. The PFA steps are as follows:

1. Parts/components and subassemblies list development

2. Identification of functions

3. Estimate the part function cost

4. Cost estimation for each component

5. Function cost modeling and total function cost calculation

6. Improvements investigation (the goal: to find high-cost parts)

7. Final cost analysis

Further development of cost reduction strategies and objectives represents a major part of the cost study. In other words, if a study is related to materials-oriented cost reduction, the objectives differ from a process-oriented study. Those two completely different types of studies indicate that a PCS knowledge subsystem is required to have all process and materials information, as well as to be openly configured to advanced information that affects cost reduction results.

Innovative change

Let us assume a component, "A," goes through the manufacturing process, which consists of milling, heat treatment, grinding, and polishing. After conducting research study, electrochemical machining has been recommended as a replacement process. This innovative process change allowed the user to improve design (shape, surface finish, tolerances) of component A, as well as to eliminate three operations. As you can see, an effective, innovative solution has been identified that led to cutting a number of sequential operations and resulted in substantial cost savings.

PART NAME	FUNCTIONS										PART COST
	A		B		C		D		E		
	%	Cost	%	Cost	%	Cost	%	Cost	%	Cost	
Filter	75	$ 0.75					15	$ 0.15	10	$ 0.10	$ 1.00
Valve			84	$ 1.05					16	$ 0.20	$ 1.25
Clip					75	$ 0.15	25	$ 0.05			$ 0.20
Bracket					80	$ 0.45			20	$ 0.11	$ 0.56
Link			82	$ 0.55			18	$ 0.12			$ 0.67
Disk	60	$ 0.36			25	$ 0.15			15	$ 0.09	$ 0.60
Cover	63	$ 0.19					37	$ 0.11			$ 0.30
Plate					33	$ 0.10			67	$ 0.20	$ 0.30
Ring							67	$ 0.10	33	$ 0.05	$ 0.15
TOTAL	26	$ 1.30	32	$ 1.60	17	$ 0.85	10	$ 0.53	15	$ 0.75	$ 5.03

Fig. 19. Hypothetical function cost form.

Over the last decade, there have been significant changes in manufacturing processes Oftentimes, new processes combined traditional operations on one piece of equipment, replacing several stations or machines. Machining centers, for example, are able to reduce cycle time and increase productivity by simultaneously removing material from several part surfaces from a single setup. It is important to realize that customers are much more able than ever before to choose from a global marketplace, and as a result, are often much more demanding in new expectations for cost. Now more than ever, managing and leveraging knowledge is a key skill, and knowledge is a vital strategic resource that needs to be developed. On the other hand, the job becomes more productive and more rewarding. When we are comparing, for example, components from different manufacturers of the same product and function, substantial cost savings (50 percent) can be found. Furthermore, if a company attributes the cost difference to factors, the following cost distribution can be presented:

Materials – 15% less

Design – 22% less

Tooling – 13%   less

The proactive cost impact assessment of design alternatives can be facilitated based on this methodology. It should be noted that CRP is the iterative type of process. In other words, when one CRP stage is completed, the cost reduction engineer begins to work on accomplishing certain objectives of the next stage of CRP. Therefore, the combined knowledge of value engineering methodology with knowledge of materials characteristics and process capabilities is very important, because cost reduction efforts on all stages require in-depth knowledge of available methods and tools. Oftentimes, an individual component of a product may come from a variety of suppliers, and may be processed several times. Each processing step may happen at a different location. An additional complication arises because of the growing interdependence between manufacturer and suppliers. This is another good example of cost avoidance identification; in other words, how to eliminate extra steps and save money. Some suggestions as to how to analyze the entire value chain should also come from a cost responsible practitioner.

The physical structure of a product has several functional subsystems that configure the product. Figure 19 allows the user to analyze the multifunctional role of a component and the contribution of each individual function, including the cost associated with it.

To remain competitive in quickly shifting markets, management continuously improves business processes and reduces product cost. A systematic approach to ongoing cost reduction process development includes: engineering contribution in terms of high-performance processes implementation, feasibility analysis, proposal cost justification, and documenting. To set up a cost reduction system, we need to establish an engineering knowledge-based subsystem together with a value analysis segment.

Value analysis includes: developing product breakdown, functional structure, creativity methods, and product performance analysis. Analysis of manufacturing technologies consists of materials, processes, equipment, and tooling applications supporting cost reduction objectives. Cost modeling, Excel spreadsheets, and modern software programs assist in finding the best cost reduction solution. All three components need to be aligned for the efficiency of cost reduction efforts.

It would be very helpful to organize continuous training related to cost reduction of a variety of manufacturing processes. First of all, it is important to develop the course package that includes:

- List of courses required

- Instruction materials

- Illustrations, charts, and tables

We would like to propose content and descriptions of courses for cost reduction.

Course title: Cost Reduction of Manufacturing Process

Course Description: The cost reduction of manufacturing process is intended to identify the range of variables that affect the process and its phases.

Overall course objectives:

- Gain an understanding of individual factors involved in reducing cost of manufacturing process

- Examine the sequential chain of operations and processing cost structure

- Develop an in-depth understanding of the investigated manufacturing process

Course content:

Lesson 1. – Introduction. Material selection, product/part design for cost reduction.

Objectives:
- Identify main properties and primary characteristics for the material that will help ensure proper material selection in the design stage for reducing cost.
- List ideas and their advantages in respect to cost savings.
- Develop a checklist recommended for cost reduction and continuous improvement.
- Establish process- and product-related measurable deliverables for cost reduction analysis.

Lesson 2. – Equipment requirements and tooling design for cost reduction.

Objectives:
- Gain knowledge in equipment requirements and correlations to cost reduction.

- Evaluate cost-saving opportunities related to tooling design and build process.
- List major drivers that mostly contribute to efficiency and productivity.

Lesson 3 – Cost reduction study: Summary and recommendations

Objectives:
- Describe results.
- Understand interrelationships between cost reduction results.
- Identify recommendations for product/part cost reduction.

The published selection of literature in support of training courses is listed in this book. This is a comprehensive set of sources that are based on technological and economical aspects of processes, and will help a company in its training process.

Recommended sources

1. *An Introduction to Efficiency and Productivity Analysis* by Tim Coelli, D.S. Prasada Rao, and George E. Battese, 1997, 296 pp.

   This book discusses several methods related to productivity analysis. The book offers a few business cases and examples, which include computer applications.

2. *Strategic Decision Making in Modern Manufacturing* by Harinder Singh Jagdev, Attracta Brennan, and J. Browne, 2003, 288 pp.

   The book examines relationships between manufacturing decision-making and business strategy at the present time. It describes investment analysis of modern manufacturing innovations.

3. *Innovations in Competitive Manufacturing* by Paul M. Swamidass, 2001, 439 pp.

   The book focuses on advances in manufacturing technology, supply chain management, and optimization. It also covers activity-based costing and target costing process.

CHAPTER 5

Recommended reading

Recommended Periodicals

Useful Internet resources

Cost estimation and cost reduction

1. *Target Costing: The Next Frontier in Strategic Cost Management* by Shahid L. Ansari, 1997, 250 pp.

   This book describes the scope, process, and tools for implementing a target costing system. The special content features: value analysis, functional cost analysis, setting prices. There are many tables and diagrams to illustrate the step-by-step activities involved in target costing.

2. *The Complete Guide to Activity-based Costing* by Michael C. O'Guin, 1991, 384 pp.

   The book presents activity-based costing system implementation process with multiple illustrations and tables. There are conceptual spreadsheets and case studies, which are helpful in understanding the ABC influence on business processes and strategies.

3. *Realistic Cost Estimating for Manufacturing: Second Edition* by William Windchell, 1989, 180 pp.

   The book contains information on specific cost models related to manufacturing processes: forging, machining, casting, welding, stamping, etc. There are many tables and diagrams with very useful numbers regarding process and materials characteristics. Many practical case studies are also presented and analyzed.

4. *Cost Estimator's Reference Manual* by Rodney D. Stewart, 1995, 717 pp.

   This is the comprehensive edition dedicated to the cost estimating process. The entire cost system is decomposed and presented in a detailed way: cost modeling, mathematical tools, material, labor, overhead cost analysis, design-to-cost process. There are chapters dedicated to parametric costing, work breakdown structure development, and computer applications.

5. *The Design of Cost Management Systems* by Robin Cooper, 1999, 536 pp.

    The book provides the extensive coverage of various cost systems with real-life practical cases. Many chapters are devoted to detailed analysis of activity-based systems in different industries. The book contains many tables, formulas, and illustrations of performance measurement methods.

6. *Practical Cost Estimating for Metal Fabrications* by George Kuprianczyk, 1996, 445 pp.

    This book presents the description of the entire costing process with multiple case studies, examples, and technical information. There are recommendations on how to improve the cost estimation system. Many pages are dedicated to detailed calculations related to cost estimates. This is the real source, useful for many practitioners assigned to cost process within a company.

7. *Design for Manufacture: Strategies, Principles, and Techniques* by John Corbett, 1991, 357 pp.

    This book is dedicated to manufacturability evaluation methods, strategies, and computer-aided techniques. There are many practical examples, tables, and diagrams, which clearly explain different case studies concerning design for manufacture analysis. Organizational activities involved in DFM process are also explained.

8. *The 60-Minute ABC Book. Activity-Based Costing for Operations Management* by Timothy S. White, 1997, 59 pp.

The book concisely provides information on activity-based costing process. This is the source for a practitioner who is just in the beginning of ABC process learning. The ABC terminology is explained. Examples and illustrations are presented.

9. *Systematic Mechanical Designing: A Cost and Management Perspective* by Mahendra S. Hundal, 1997, 561pp.

The book presents the systematic approach to design process and product cost assessment. There are chapters dedicated to design improvements analysis, cost estimation of a variety of different technologies and products. Materials and manufacturing processes information is presented in tables, and diagrams as well. Economical process selection is explained with use of mathematical formulas, which makes the information especially helpful. The design for cost is presented with practical examples and illustrations. This is a good source for engineers and managers.

10. *CASA/SME. Blue Book: Cost engineering: The practice and the future*, by Rajkar Roy, PhD, Peter Sackett, PhD, 2003, 19pp.

This book describes cost engineering concepts, current status. Cost reduction approaches are discussed and modern practices presented. The focus of this work is on a variety of industries.

11. *Manufacturing Cost Engineering Handbook,* by Eric M. Malstrom, 1984, 447pp.

Manufacturing cost development and analysis are presented. Methods of cost elements calculations are described with examples. The attention to details makes this book useful for cost estimators as well as for specialists with cost estimating responsibilities and assignments.

12. *Estimating and Costing for the Metal Manufacturing Industries* by Robert C. Creese, M. Adithan, 1992, 270 pp.

This book is a good source for cost studies. The main topics are presented as follows: conceptual cost estimating techniques, material costing, overhead cost analysis, basic costing for machining, welding, casting, forging.

13. *Target Costing and Value Engineering* by Robin Coper and Regine Slagmulder, 1997, 379 pp.

This research aims to explore forms of cost modeling. Target costing is explained and analyzed from different aspects. The book is illustrated with many diagrams, tables, and charts, which enables the user to visualize the role of individual elements of target cost analysis. Examples from the automotive industry are presented.

14. *Value Engineering, Analysis and Methodology* by Del L. Younker, 2003, 326 pp.

This book is about methods for engineering alternatives analysis, cost of functions, creativity and developing new ideas, design concepts, and problem solving. Life cycle cost components analysis is described. This is the integrated source of value analysis and cost evaluation methods.

15. *Implementing Concurrent Engineering in Small Companies* by Susan Carlson Skalak, 2002, 306 pp.

The benefits of concurrent engineering (CE) are presented. This book analyzes the manufacturing modeling and design process in the light of CE. Author explains risk assessment modeling and product targets development. Design examples of sheet metal and plastic components are presented. Alternative concept cost comparison is described.

16. *Value Engineering* by Richard Park, 1998, 340 pp.

The book is dedicated to value engineering. It contains cost estimating methodology and techniques, financial tools, payback analysis, graphical information, and value engineering methods and application examples, enabling the user to visualize the advantages of value analysis. There are mathematical formulas and tables.

17. *Profit-Focused Supplier Management: How to Identify Risks and Recognize Opportunities* by Pirkko Ostring, 2003, 256 pp.

This book is about managing relationships with suppliers. There are many recommendations about continuous improvements, identifying opportunities related to purchasing activities with suppliers.

18. *Activity-Based Costing: Problems in Practice* by Ian Cobb and John Innes, 1992, 36 pp.

The book is about activity-based costing (ABC) implementation. There are helpful practical examples for professionals as well as for management related to ABC. This is a good source for cost estimators, engineers, and purchasing specialists.

19. *Pricing for Profitability: Activity-Based Pricing for Competitive Advantage* by John L. Daly, 2001, 288 pp.

This is a complete source for pricing strategies, activity-based pricing, target costing, and price negotiations. In addition, the book examines the influence of capacity utilization.

Modern Manufacturing Technology: Processes, operations, tooling

1. *Computer-Aided Injected Mold Design and Manufacture* by Jerry Y.H. Fuh and Y.F. Zhang, 2004, 360 pp.

This book brings together mold design process, assembly, and software modeling associated with die and mold manufacturing. Computer-aided process planning in mold making is analyzed. There is a chapter dedicated to cost estimation of injection molds.

2. *Product Design and Development* by Karl T. Ulrich, 2000, 359 pp.

The book contains information on product development process, cost studies, industrial examples, and cost reduction analysis. The product concept development with cost considerations is also presented. Material and manufacturing processes are examined in terms of selection and implementation in production. It is a good source for cross-functional teams involved in product development and analysis.

3. *Advanced Machining Technology Handbook* by James Brown, 1998, 580 pp.

The comprehensive source on major categories of manufacturing technology with in-depth analysis of equipment technical parameters and materials. Casting, forging, traditional and nontraditional machining, as well as industrial finishing processes and systems are illustrated with pictures, tables, and diagrams, enabling the user to gain knowledge in contemporary improvement methods and apply it to real daily practical situations. It is like the engineering library on your desk. In addition to that, all technologies are described from their economic aspects, advantages, and limitations.

4. *Design for X. Concurrent Engineering Imperatives* by G.Q. Huang, 1996, 489 pp.

   This book is intended to analyze design for competition, design for manufacturing and assembly, design for reliability, and product cost evaluation associated with these design for X methods. Multiple practical cases help to gain knowledge of functional cost analysis and apply it to cost reduction strategies development.

5. *Exploring Advanced Manufacturing Technologies* by Steve Krar and Arthur Gill, 2003.

   The book is intended to provide the effective source of the major manufacturing processes and machines including automation systems. Among described processes: casting, machining, near-net-shaping technology, nontraditional machining processes, and their parameters. This a good reference source for better process selection.

6. *Processes and Design for Manufacturing* by Sherif D. El Wakil, 2002, 614 pp.

   The book provides practical information about a variety of manufacturing processes, such as: powder metallurgy, forming, machining, and materials (metals, plastics). A special chapter is dedicated to product cost estimation.

7. *Applied Manufacturing Process Planning: with emphasis on metal forming and machining* by Donald H. Nelson, 2000, 720 pp.

   The book is intended to analyze the process and machine selection for manufacturing metal components. There are chapters dedicated to tolerance charting, machining operations, and workpiece holding techniques. Among representatives of manufactured components are: machined parts, upset parts, sheet metal parts. The book would be very helpful for engineering, purchasing, and managers.

8. *Hybrid Machining Process Evaluation and Development* by Jerzy Kozak and Kamalakar P. Rajurkar, 1999, 125pp.

Advanced manufacturing processes are examined. New applications and technological developments in machining methods and their parameters are described. There are many tables and illustrations of combined manufacturing processes.

Recommended periodicals

1. *Modern Applications News* (MAN)

2. Mechanical Engineering

3. Metalforming

4. Stamping Journal

5. Fabrication and Welding

6. American Machinist

7. Moldmaking Technology

8. *Automotive Engineering*

Internet resources

1. www.Engineering.com

2. www.Advancededm.com

3. www.Edmstore.com

4. www.Gensparkamerica.com

5. www.matweb.com

6. www.edmtechnologies.net

SUMMARY AND FINAL COMMENTS

Advanced processes offer a radically different way of approaching the production process. There are several important considerations that arise when dealing with new technology. This book covers in an integrated fashion the complete route from modern manufacturing through engineering knowledge systems (EKS) and cost evaluation to cost reduction analysis. Expanded knowledge enriches positive thought processes, links ideas in a very efficient way, and improves the professional's delivery skills. Today, a broader view should be considered by practitioners when planning the production process or developing manufacturing strategies. Furthermore, nontraditional technologies can be applied with traditional methods while reducing cost and improving product quality. Also, such combined processes are very effective in making dies and molds from hard-to-machine and heat-treated materials. For example, electrochemical machining can be successfully applied to dies for powder metallurgy that requires high mechanical properties (hardness). Modern machines and presses for near-net-shape methods are able to utilize computer systems and control the entire process from start to finish. Engineering knowledge systems (EKS) became a very important part of the production process. EKS contains materials, machines, processes, and advanced applications. It enables the user to synthesize the new operational process, applying creativity, value engineering, and value analysis. EKS includes technological and engineering developments, which significantly contributes to optimized method selection based on best practices. This is what increases product competitiveness. Today, like never before, the progress in manufacturing cannot be achieved without advanced and well-maintained EKS.

Product cost assessment should be organized with systematic method and engineering vision as follows: process planning tools, materials cost calculations, detailed burden analysis, manufacturing cost equations, and computer cost models. Cost studies based on system provide support of financial performance analysis, business planning, and development. They also help to find cost reduction opportunities and conduct cost benefit analysis.

Modern manufacturing technology and cost analysis enables users to identify the following factors:

- The growing demand for higher quality and functionality put enormous strain on management.

- Moving to nontraditional processes as well as to near-net-shape technology, stimulates the inventiveness, creativity, and innovations in manufacturing environments.

- Product competitiveness can be achieved by research and implementation of modern technologies, machines, and tooling based on new technological and engineering developments.

- The useful applications of physics and chemistry with associated principles leads to nontraditional technologies and product designs.

- Modern types of materials applied by designers require a systematic approach to the selection of forming process, machines, presses, and tooling.

- Design for manufacturing based on nontraditional technological applications improves product functions and features, as well as its parameters, which cannot be achieved with traditional technology.

- Continuous education is a lifelong process; professionals with adequate technical knowledge and training are able to substantially improve product, reduce cost, and increase productivity.

- The modern enterprise operating models require nontraditional processes, providing high flexibility in terms of quick response to changes in product lines and manufacturing.

APPENDIXES

Templates, Forms, Tables

COST ITEM	SPECIFICATION	RATE	USAGE	UNITS OF MEASURE	TOTAL COST	NOTES
MATERIALS 1. Material A 2. Material B 3. Purchase Part A 4. Purchase Part B 5. Other						
LABOR 1. Fabrication Processes 2. Assembly Processes 3. Test 4. Inspection 5. Other						
BURDEN 1. Equipment Cost 2. Depreciation 3. Facilities 4. Insurance						
5. Utilities 6. Maintenance 7. Other						

Total Product Cost

Appendix 1.

#	DESCRIPTION	CURRENT			PROPOSED			VARIANCE	
		MATERIAL COST	LABOR COST plus OVERHEAD	MANUFACTURING COST	MATERIAL COST	LABOR COST plus OVERHEAD	MANUFACTURING COST	$	%
	FLANGE								
	COVER								
	DISK								
	BRACKET								
	PLATE								
	BOLT								
	NUT								
	WASHER								
	ASSEMBLY								
	TOTAL								

Appendix 2

114

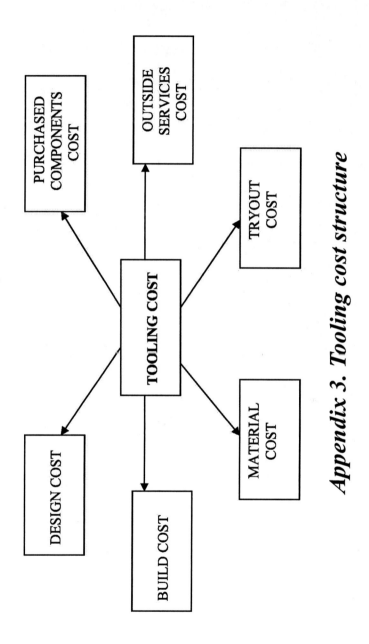

Appendix 3. Tooling cost structure

OPERATION	MACHINE TOOLS APPLIED
TURNING	LATHE, BORING MILL, VERTICAL SHAPER, MILLING MACHINE
DRILLING	DRILL PRESS, MACHINING CENTER, VERTICAL MILLING MACHINE, LATHE, HORIZONTAL BORING MACHINE, HORIZONTAL MILLING MACHINE
REAMING	LATHE, DRILL PRESS, BORING MILL, HORIZONTAL BORING MACHINE, MACHINING CENTER, MILLING MACHINE
BORING	LATHE, BORING MILL, HORIZONTAL BORING MACHINE, MACHINING CENTER, MILLING MACHINE, DRILL PRESS
GRINDING	CYLINDRICAL GRINDER, LATHE (WITH SPECIAL ATTACHMENT)
MILLING	MILLING MACHINE, MACHINING CENTER, LATHE (WITH SPECIAL ATTACHMENT), DRILL PRESS (LIGHT CUTS)
FACING	LATHE, BORING MILL
BROACHING	BROACHING MACHINE
SAWING (OF PLATES)	CONTOUR OR BAND SAW, FLAME CUTTING OR PLASMA ARC
BROACHING	BROACHING MACHINE, ARBOR PRESS (KEYWAY BROACHING)
SHAPING	HORIZONTAL SHAPER, VERTICAL SHAPER
PLANING	PLANER

Appendix 4 Operations and applied machine tools

TIME CONVERSION TABLE

Seconds	Minutes	Hours	PCS/Hour
1	0.017	0.0003	3600
2	0.033	0.0006	1800
3	0.050	0.0008	1200
4	0.067	0.0011	900
5	0.083	0.0014	720
10	0.167	0.0028	360
15	0.250	0.0042	240
20	0.333	0.0056	180
30	0.500	0.0083	120
40	0.667	0.0111	90
50	0.833	0.0139	72
60	1	0.0167	60
90	1.500	0.0250	40
120	2	0.0333	30
150	2.500	0.0417	24
180	3	0.0500	20
210	3.500	0.0583	17.1
240	4	0.0667	15
270	4.500	0.0750	13.3
300	5	0.0833	12
600	10	0.1667	6
900	15	0.2500	4
1200	20	0.3333	3
1500	25	0.4167	2.4
1800	30	0.5000	2
2400	40	0.6667	1.5
3000	50	0.8333	1.2
3600	60	1.0000	1

Appendix 5. Time conversion table.

#	PART NAME	CURRENT					PROPOSED					VARIANCE	
		INJECTION MOLDING	FORGING	CASTING	MACHINING		INJECTION MOLDING	FORGING	CASTING	MACHINING		$	%
	COVER												
	DISK												
	PLUG												
	FLANGE												
	CARRIER												
	GASKET												
	BRACKET												
	BAR												
	GUIDE												
	TOTAL												

Appendix 6. Conceptual form for manufacturing cost analysis

#	PART NAME	PART #	QUANTITY	MATERIAL COST	LABOR COST	MACHINE COST	MANUFACTURING COST	EXTENDED COST
	KNOB							
	PLATE							
	RING							
	ANGLE							
	BRACKET							
	CLAMP							
	CLIP							
	CYLINDER							
	VALVE							
	COVER							
	FASTENERS							
	TOTAL							

Appendix 7. Hypothetical example of manufacturing cost worksheet

Appendix 8. Cycle Time Conversions

Cycle Time (number of seconds)	Parts Per Hour Based On Cycle Time and Efficiency				
	Gross Production	90% Efficiency	80% Efficiency	70% Efficiency	60% Efficiency
1	3600	3240	2880	2520	2160
1.25	2888	2600	2315	2022	1732
1.5	2400	2160	1930	1680	1441
1.75	2056	1849	1643	1440	1233
2	1800	1620	1440	1263	1082
2.5	1440	1296	1152	1008	864
3	1200	1080	960	840	720
3.5	1029	925	822	719	616
4	900	810	720	630	540
4.5	800	720	640	560	480
5	720	648	576	504	432
5.5	654	589	523	458	393
6	600	540	480	420	360
6.5	554	498	443	388	332
7	514	463	412	360	309
7.5	480	432	384	336	288
8	450	405	360	315	270
8.5	423	381	339	296	254
9	400	360	320	280	240
9.5	379	341	303	265	227
10	360	324	288	252	216
11	327	295	262	229	196
12	300	270	240	210	180
13	277	249	221	194	166
14	257	231	206	180	154
15	240	216	192	168	144
16	225	202	180	157	135
17	212	190	169	148	127
18	200	180	160	140	120
19	189	170	151	132	114
20	180	162	144	126	108

Cycle Time (number of seconds)	Parts Per Hour Based On Cycle Time and Efficiency				
	Gross Production	90% Efficiency	80% Efficiency	70% Efficiency	60% Efficiency
21	171	154	137	120	103
22	163	147	131	114	98
23	156	141	125	110	94
24	150	135	120	105	90
25	144	129	115	101	86
26	138	124	110	97	83
27	133	120	107	93	80
28	128	115	103	90	77
29	124	111	99	87	74
30	120	108	96	84	72
35	103	92	82	72	62
40	90	81	72	63	54
45	80	72	64	56	48
50	72	65	57	50	43
55	65	59	52	46	39
60	60	54	48	42	36
70	51	46	41	36	31
80	45	40	36	31	27
90	40	36	32	28	24
100	36	32	29	25	22
110	33	29	26	23	19.6
120	30	27	24	21	18
140	26	23	20	18	15.4
160	22	20.2	18	15.7	13.5
180	20	18	16	14	12
200	18	16.2	14.4	12.6	10.1
220	16.4	14.6	13.1	11.5	9.8
240	15	13.5	12	10.5	9
260	13.8	12.5	11.1	9.7	8.3
280	12.8	11.5	10.3	9	7.7
300	12	10.8	9.6	8.4	7.2

Appendix 9.

American Society for Materials Information
    ASM International
    9639 Kinsman Road
    Materials Park, OH 44073-0002 USA
Phone:
    440-338-5151
    800-336-5152 (U.S. and Canada)
    800-368-9800 (Europe)
Fax:
    440-338-4634
E-mail:
    ASM Customer Service Center: *cust-srv@asminternational.org*

The American Society of Mechanical Engineers
    ASME International
    Three Park Avenue
    New York, NY 10016-5990
Phone:
    800-843-2763 (U.S./Canada)
    95-800-843-2763 (Mexico)
URL:
    http://www.asme.org
Email:
    infocentral@asme.org

American Society for Testing and Materials (ASTM)
    ASTM
    100 Barr Harbor Drive
    West Conshohocken, PA 19428-2959
Phone:
    (610) 832-9585
Fax:
    (610) 832-9555
URL:
    http://www.astm.org

Society of Manufacturing Engineers
　One SME Drive
　Dearborn, MI 48121
Phone:
　800-733-4763
　313-271-1500
Fax:
　313-425-3401
URL:
　http://www.sme.org

SAE World Headquarters
　400 Commonwealth Dr.
　Warrendale, PA 15096-0001 USA
URL:
　http://www.sae.org

American Purchasing Society, Inc
　North Island Center, Ste. 203
　8 East Galena Blvd.
　Aurora, IL 60506
Phone:
　630-859-0250
URL:
　http://www.american-purchasing.com
E-mail:
　popurch@mqci.com

Institute of Industrial Engineers
　3577 Parkway Lane, Suite 200
　Norcross, GA 30092
Phone:
　800-494-0460
770-449-0460
Fax:
　770-441-3295
URL:
　http://www.iienet.org

APICS The Association for Operations Management
 5301 Shawnee Road
 Alexandria, VA 22312-2317
Phone:
 800-444-2742
 703-354-8851
 Fax:
 703-354-8106
URL:
 http://www.apics.org

Fabricators & Manufacturers Association, International (FMA)
 833 Featherstone Rd.
 Rockford, IL 61107, USA
Phone:
 815-399-8775
Fax:
 815-484-7701
URL:
 http://www.fmanet.org
E-mail:
 info@fmanet.org

AACE International (Association for the Advancement of Cost Engineering)
 209 Prairie Ave., Suite 100,
 Morgantown,WV 26501, USA
Phone:
 304-296-8444
URL:
 http://www.aacei.org

Tooling & Manufacturing Association (TMA)
 1177 S. Dee Road
 Park Ridge, IL 60068
Phone:
 847-825-1120
URL:
 http://www.tmanet.com

National Society of Professional Engineers (NSPE)
    1420 King Street
    Alexandria, VA 22314-2794
Phone:
    703-684-2800
Fax:
    703-836-4875
URL:
    http://www.nspe.org

North American Die Casting Association (NADCA)
    241 Holbrook Dr
    Wheeling, Illinois 60090-5809, USA
Phone:
    847-279-0001
Fax:
    847-279-0002
URL:
    http://www.diecasting.org

American Foundry Society (AFS)
    1695 N Penny Lane
    Schaumburg, IL 60173
Phone:
    800-537-4237
URL:
    http://www.afsinc.org

About the Author

Michael Lembersky has over twenty five years of experience and educational background in production engineering and cost estimation. He holds a M.S. degree in Mechanical Engineering with concentration in machining and forming processes. He began his career as manufacturing engineer developing machining, forming technology (stamping, casting, plastic molding) and non-traditional (electrical discharge machining, electrochemical machining, etc.) applications for a variety of product systems resulted in significant cost reduction. Michael conducted research and scientific projects in modern technology; identified and evaluated technological and economical trends in manufacturing environment. He worked for medium and large size industrial companies. He constantly applied educational background and experience in variety of processes such as mechanical machining, sheet metal stamping, electrochemical machining, die casting, plastic technology to achieve productivity increase and significant dollar savings by replacing materials, eliminating loses in production. He always helped companies to benefit from selection of different technologies. Michael contributed to success of cross-functional teams by discovering opportunities for product innovations, which led to sufficient economical achievements and reduced time-to-market. He provided complex cost assessment studies and competitive analysis resulted in multiple components reduction and product performance improvements. Michael developed engineering knowledge databases, conducted value engineering projects and implemented quality assurance systems. Managed research projects related to industrial components and their molds and dies, which resulted in million dollars savings. He has continuously expanded knowledge and expertise by completing courses in Value Engineering /Value Analysis, Design for Manufacturing, Tooling Cost Estimation, SAP Implementation. Gained extensive experience working with Enterprise Resource Planning (ERP) systems and advanced costing software. Michael organized cost systems and generated economical models which served as tools for more accurately and efficiently analyzing proposals, new product concepts and design modifications to existing products. He constantly applied his comprehensive experience and skills to support effective make or buy decisions and supplier evaluation. He provided assistance to colleagues and developed training instructions related to advanced machining and metalforming applications. His research interests focus on technical innovations, feasibility analysis and economical justification of advanced processes. Michael is a Member of Society for Manufacturing Engineers (SME) and Society for Automotive Engineers (SAE).

Lana Lembersky has extensive experience in development of financial and cost applications. She gained valuable expertise in knowledge systems and manufacturing proposal cost analysis. She holds a M.S. degree in Computer Science. Her many years experience as application developer give Lana Lembersky the opportunity of sharing her knowledge with practitioners. She specialized in complex cost evaluation applications, developed several systems ranging in size from individual applications utilizing database-management systems to enterprise-wide client/server applications.

Natalya Lembersky is a graphic designer. She holds a B.S. degree in Fine Arts. Her illustrations had earned awards from the Detroit Society of Professional Journalists. She has wide experience working with magazines, agencies, businesses and Web sites. Member of the Graphic Artists Guild.

Printed in the United States
62441LVS00003B/46

9 781420 868708